Designing Displays
for Older Adults

Human Factors & Aging Series

Series Editors
Wendy A. Rogers and Arthur D. Fisk
School of Psychology
Georgia Institute of Technology – Atlanta, Georgia

Human Factors
& Aging Series

Designing Displays
for Older Adults

Richard Pak
Anne McLaughlin

CRC Press
Taylor & Francis Group
Boca Raton   London   New York

CRC Press is an imprint of the
Taylor & Francis Group, an **informa** business

CRC Press
Taylor & Francis Group
6000 Broken Sound Parkway NW, Suite 300
Boca Raton, FL 33487-2742

© 2011 by Taylor and Francis Group, LLC
CRC Press is an imprint of Taylor & Francis Group, an Informa business

Printed in the United States of America on acid-free paper
10 9 8 7 6 5 4 3 2 1

International Standard Book Number: 978-1-4398-0139-0 (Paperback)

Library of Congress Cataloging-in-Publication Data

Pak, Richard.
  Designing displays for older adults / Richard Pak, Anne McLaughlin.
    p. cm. -- (Human factors and aging series)
  Includes bibliographical references and index.
  ISBN 978-1-4398-0139-0 (pbk. : alk. paper)
  1. Human engineering. 2. Older people. I. McLaughlin, Anne. II. Title. III. Series.

TA166.P34 2011
620.8'2--dc22   2010026997

Visit the Taylor & Francis Web site at
http://www.taylorandfrancis.com

and the CRC Press Web site at
http://www.crcpress.com

Contents

Preface

This book is focused on the design of displays for the older user. Why does this topic deserve a book? Aging leads to a complex set of changes, both mentally and physically, that can affect technology acceptance, adoption, interaction, safety, and satisfaction. Design with an understanding of these changes will result in better products and systems for users in all stages of the lifespan. Conventional wisdom (possibly informed by personal experience) is that getting older leads to a decline across a broad set of skills and abilities; however, the reality is that as some capabilities decline with age, others remain stable or increase. For example, although a 60-year-old man may not be able to beat his granddaughter in the computer puzzle game Tetris, the elder will invariably beat the youth in games of knowledge such as Trivial Pursuit or the television quiz show *Jeopardy*. Design of displays and technology can capitalize on these capabilities to ameliorate the limitations that can come with age.

As human factors professionals, we have often been frustrated at how little research makes it to practice. This is why the target audience for this book is a usability engineer, or user interface/user experience designer who is tasked with creating an interface that might be used by older adults. Literally hundreds of papers have been written about interface issues experienced by older adults, but how many actually influence the designs older adults use? We believe the challenge comes in part from the sheer number of articles available. Design and usability evaluation are fast-paced activities with little time allowed for literature review. Many professionals do not have the time to sift through thousands of papers to determine (a) which are related to the question at hand and (b) whether the design or study has merit. Another reason may be that academic papers typically target other academics and may not stress the application or design implications of their findings. Finally, another barrier to knowledge transfer may be that academic publishing moves slower than the design and usability industry. The time it takes for journal articles to reach the audience, from submission time, can be months to years. Because of this, authors may be loathe to nail down concrete design guidelines, opting for the conceptual and general (but often vague and hard to implement), because their research may be published a year in the future. This book distills decades of published aging research most relevant to the design of displays.

We believe this book offers a benefit beyond individual research studies. The first half of the book is a primer of age-related changes in cognition, perception, and behavior. Theory can be used to organize examples from the literature into meaningful principles that improve understanding. Using theory backed up by evidence provides an understanding of *why* we see certain problems with many displays and often predicts solutions. This understanding surpasses an individual interface and provides the practitioner with ways to plan for older users on multiple display types. We then apply these theories in real design exercises. In all chapters we provide specific guidelines for display examples to bridge theory and practice.

Author Note

Author order for this book was determined randomly and represents equal knowledge and work from the authors. Dr. Richard Pak is currently an assistant professor of psychology and directs the Cognition, Aging, and Technology lab at Clemson University. Dr. Anne Collins McLaughlin is currently an assistant professor of psychology, directs the Learning, Aging, and Cognitive Ergonomics lab, and codirects the Gains Through Gaming lab, both at North Carolina State University. We thank the series editors, Wendy Rogers and Dan Fisk, for their comments and suggestions. We would also like to thank the following individuals for taking the time to review individual chapters: C. Travis Bowles, Michelle Bryant, Doug Gillan, Nicole Fink, Mary Luong, Jeremy Mendel, Margaux Price, Kristen Miller, Rick Tyrrell, John Sprufera, and Ray Stanley.

chapter one

Introduction

Age is an indicator of how long one has lived but is not a complete indicator of a specific individual's capabilities and limitations. One can easily imagine how a physically fit 55-year-old triathlete could outperform a 34-year-old in a marathon, personified by Cliff Young, the 61-year-old winner of the 1983 Sydney ultramarathon (544 miles). Similarly, with respect to cognitive capacity, there is wide variety in capabilities and limitations that are linked with age. Thus the definition of "old" and "older" can be a tricky issue. The issue is further complicated by the sheer variability in any given ability as we get older. Generally, for younger and middle age groups, capabilities vary, but this variability widens as people get older; the older adult age group (defined as those aged 65 and over) are more different from each other than people in other age groups differ from persons in their own age group.

One way to think about the older user is via his or her familiarity with current technologies and interface conventions. Jakob Nielsen reported that "Between the ages of 25 and 60, the time users need to complete website tasks increases by 0.8% per year." But does this time increase come from a rapidly changing technology or declines in the human body? Both culture and the physical aging process play a role. Such dual causes for the same symptom exemplify why designs should be carefully analyzed as to the challenges they produce. It is important to know how much of a role the display plays in increased time to complete tasks versus slowed completion times due to the user's inexperience with that type of task, or any number of other variables.

Another way to evaluate the older user is by the appearance of perceptual and cognitive changes that we usually associate with aging. This may include farsightedness, the need for bifocals or reading glasses, hearing aids, and an increased reliance on notes rather than memory for everyday tasks. An understanding of these changes and the effects they can have on display use is critical, because these changes often interact with each other. Thus, understanding a single age-related change, such as vision, can lead to designs that adversely affect other senses and cognitions, such as the high working memory requirements of audio displays or creating the need to scroll larger text, which can require precise movement.

1.1   What do older adults want from technology? What do they do with technology?

Across a wide range of everyday activities, users encounter electronic technology. Technology, for the purposes of this book, is broadly defined as any tool or artifact that helps the user accomplish a task, limited to electronic displays for the purposes of this book. Even with this limitation, consider how ubiquitous technology is for most people. Table 1.1 shows everyday technologies located in typical environments.

Many of these technologies are either specifically for or related to communication with others. Mobile communication occurs commonly with all age groups, though use of mobile communication technologies can vary by age (Table 1.2). Those over age 65 are less active users of the full range of advanced mobile services, but they are enthusiastic users of mobile voice communications, especially in emergency situations (Table 1.3). The need to communicate is enabled through cell phone use, and thus the product has been adopted by older adults. Indeed, this adoption occurred in spite of the fact that many cell phones are not well designed for older users. These design problems can partly explain the lower use in older age groups compared to younger. When a user group is excluded from these everyday technologies by designs that do not accommodate them, both the quality of life of those users and the market share of the product companies can suffer.

Similar to cell phones, use of the internet as a tool beyond information seeking is becoming more common among older adults. According to a 2006 Pew survey, 41% of internet users are over age 65. Online banking is a popular activity, and 43% of internet users engage in some form of online banking. However, only 27% of users aged 65 or older regularly banked online. This is a fairly typical finding when relating age and technology use, but there are many potential reasons as to why. These reasons range from usability issues with the interface to unease in accessing financial information over the web (mistrust). These are two different reasons for avoiding online banking, and one would take different paths to overcome them. The lesson from investigating reasons for nonadoption of a technology is that it is important not to conclude that older adults avoid technology for any stereotypical reason, as avoidance is often affected by context, needs, and experience levels. Understanding the barriers to the adoption of potentially useful services and products is crucial to overcoming the problems and increasing adoption. Stereotypes of older users should be avoided in favor of evidence-based analyses.

Table 1.1 Everyday Technologies in Different Environs

• In the home	• Heating and cooling controls
	• Kitchen appliances
	• Internet-based banking on computer
	• Personal medical devices
	• Television news, weather
	• DVD players
	• Game consoles
	• Cable set-top boxes
• In the car	• Navigation systems
	• Trip computer
	• Mobile phone
	• Climate controls
	• Speedometer
• Everyday errands	• Automated teller machines
	• Parking meters
	• Point of sale terminals
• Traveling	• Ticket kiosks
	• Arrival/departure screens at airport
	• E-mail reminders of delays
• At work	• Word processors
	• Advanced copying/multifunction machines
	• Teleconferencing equipment
	• Phone menu systems
	• Domain-specific technologies such as machines
• General communication	• Answering machines and voice-mail
	• Cordless phones
	• Mobile phones
	• Smart phones
	• Pagers
	• Computers

1.2   Stereotypes of older users

A common stereotype of older adults is that they do not and will not use technology. If this were true, there would be no need for this book; all displays and interfaces would be translated to disinterested older adults by their children and grandchildren. However, this stereotype could not be farther from the truth. Adults over 65 *want* to keep up with technology

Table 1.2 Wireless/Mobile Activities by Age

Those who have a cell phone or personal data assistant who have ever done one of listed activities				
	18–29	30–49	50–64	65+
Send or receive text messages	92%	76%	50%	17%
Take a picture	87	71	59	29
Play a game	46	32	12	6
Send or receive mail	34	30	17	7
Access the internet	39	31	14	4
Record a video	32	21	11	2
Play music	43	21	7	5
Send or receive instant messages	34	21	12	7
Get a map or directions to another location	27	24	11	5
Watch video	24	15	7	3
Percent who have done at least one of these activities	93	80	59	27
Median number of activities ever done	4	2	1	0
Number of cases	296	578	506	399

Source: Pew Internet & American Life conducted from March 26 to April 19, 2009. N = 2,253. Margin of error is ±/−2. Survey conducted in English. Horrigan, J. (2009). Wireless Internet Use. http://www.pewinternet.org/Reports/2009/12-Wireless-Internet-Use.aspx, access on September 10, 2009. With permission.

and take advantage of what a technological world has to offer. About half of persons aged 65 to 74 are cell phone subscribers, and one-third over the age of 75 pay for service. The Center for the Digital Future found that in 2009, 40% of persons over 65 in the United States were internet users. Participants in our research studies frequently mention that understanding new technologies makes them feel connected to others and the world in general.

Use of the internet is one microcosm of older adults' perception of technology. Though the statistic of 40% using the internet seems impressive, it is paltry when compared to the nearly 100% of younger users who take advantage of the web on a daily basis. A common stereotype of older users is that they are unable to learn to use complicated technological systems. However, when older adults reject technology it tends to be due to *not perceiving a benefit* of the technology, not necessarily because it is too difficult or time-consuming to learn. The end result may be the same, fewer older adults use new technologies, but the reason is important. When older adults perceive a benefit, they are willing to invest the time

Table 1.3 Cell Phone Features Utilized by Different Age Groups

Percentage of cell phone owners in each age cohort who say ...	Age 18–29	Age 30–49	Age 50–64	Age 65+	Total
I personalized my cell by changing wallpaper or adding ring tones	85	72	50	29	65
I have used my cell phone in an emergency, and it really helped	79	76	70	65	74
I often make cell phone calls to fill up my free time while I'm traveling or waiting for someone	61	43	25	20	41
I have occasionally been shocked at the size of my monthly cell phone bill	47	38	26	23	36
When I'm on my cell phone, I'm not always truthful about exactly where I am	39	23	9	10	22
Too many people try to get in touch with me because they know I have a cell phone	37	23	12	5	22
I often feel like I have to answer my cell phone even when it interrupts a meeting or meal	31	26	14	20	24
I received unsolicited commercial text messages	28	15	15	13	18
I have drawn criticism or dirty looks because of the way I use my cell in public	14	9	4	3	8
I used cell to vote in a contest shown on TV such as "American Idol"	14	9	5	1	8

Source: Ranie, Lee. Americans and their cell phones. Pew Internet & American Life project, April 2006. With permission.

to learn. However, an unusable interface is more likely to tilt the scale in favor of "not worth it."

E-mail provides another useful example to illustrate these points. E-mail is a form of communication, both business and personal. Imagine someone having only knowing e-mail is for communication. Would one understand that e-mail allows instant communication? Would one know that they could send and receive pictures of the people they care about the very day the pictures were taken? Would they know that e-mail is free? Would they understand the asynchronous nature of e-mail; that the person they are communicating with did not have to be available at the instant the e-mail was sent, but that the message would be there waiting for the recipient(s), or even that the same message could be sent to more than one person at the same time? If potential users do not know these things, there is no reason to prefer e-mail over a letter or a phone call. It should not be assumed that "everyone" understands these benefits of e-mail, and if a person does not know of these benefits there is little reason to adopt the technology.

1.3   Universal design

Universal design is a popular paradigm for the goals of any design process. In short, universal design principles advocate there are ways to design displays (and more generally, all products) in ways that make them usable for persons with a disability *and* persons with no disability. The biggest problem with universal design is the circularity of the definition: If something makes a product easier to use, faster to use, and more pleasant to use by all people, then it is universally designed. If it fails in any of those categories, it is not universally designed. Being better for all users is a lofty goal, and in some instances universal design succeeds in improving a display for all persons. One example is the pattern of raised bumps on a sidewalk and darker concrete mix that signal a ramp to the street and a pedestrian crossing. The bumps do not inhibit able users, but provide an important signal to the visually impaired and cane or wheelchair users.

Unfortunately, there are instances where increasing usability for some may reduce usability for others. For example, making an interface usable for people with visual impairments could involve forcing unnecessary audio on all users. Other users may still be able to use the display with included audio, but it may have other deleterious effects such as slowing or frustrating the user. Thus, we recommend thinking of universal design as a philosophy: Many times there are ways to improve designs so they may be more easily used by everyone, but sometimes having an adaptive or adaptable interface can provide more assistance than any single design. This book provides a thorough overview of designing such adaptations for an older population.

1.4   What is a display?

We are inundated with computerized displays to the point that it can be difficult to notice all of them. For the purposes of this book, we define a display as any electronic device that presents text or graphics for the purpose of informing the use of possible actions or the effects of their actions. The term is synonymous with *user interface (UI)* and will be used interchangeably. This definition is purposely broad to include a range of things encountered on a daily basis (Figure 1.1). Many displays are primarily visual with some auditory components (e.g., warning tones and beeps), but some displays may be entirely auditory, such as a

Figure 1.1 Everyday displays.

voice-activated telephone menu. Sound can be both feedback from a display and part of the display itself. For example, a computerized beep tells the ATM user that a button was pressed, whereas a telephone menu system is an audio display: It provides all the information needed to make input decisions. Still others are primarily movement based, such as a computerized sewing machine or video game systems like the Nintendo Wii. Displays present information with the purpose of helping the user make some kind of decision and then act. In order to act, the user would input information using some kind of input device. These include input devices, sound, and sometimes tactile information. Virtually every display has a combination of visual, auditory, and movement demands to understand or operate it.

1.4.1 Input devices

For some displays the input device is physically the same as the display, as in the case of a touch screen. However, in many displays, the input device is physically separate from the visual display, as in the case of a computer mouse. The input device is still considered part of the display and is often a display in itself. One example would be a TV remote control or a cell phone keypad. How the keypad is designed affects interaction with the screen display. The link between input devices and displays is the reason we discuss input device design across many chapters in this book.

1.5 Goals for the book

The purpose of this book is to provide information about the fundamental changes in perception and cognition that tend to come with aging and to directly link this information to their effect on the use of displays and interfaces by older users. We point out aspects of displays that may be sensitive to age-related differences in perception (vision and hearing), cognition, and movement ability and suggest ways to compensate for such age-related changes. In many cases subtle changes to a display can make it much more usable for older users, and in some cases more drastic changes are necessary. We believe a fundamental understanding of age-related change is the most important information we can provide and that such information can influence design from the very beginning stages, rather than waiting for testing to reveal the problems users have with the product. Our examples are meant to demonstrate how commonly these issues appear in displays, while offering specific advice on ways to detect and avoid these pitfalls.

One of the reasons there can be so many age-related issues with a display is the inherent complexity of technology. Table 1.4 contains a task analysis of web browser functions illustrating the constellation of abilities required for interacting with a simple display. Without using tools such as task analysis, displays can appear to be simpler than they actually are. One famous example comes from a home medical device (a blood glucose meter) that was advertised as three easy steps. A task analysis revealed over 50 steps to complete a task with the device.

1.6 Accessibility guidelines

In addition to the conceptual information we offer on aging, there are a number of resources designers may use to ensure displays are aging friendly. Note that having a display that complies with aging guidelines or the Americans with Disabilities Act does *not* guarantee a usable display. It is still important to understand conceptually what older users may be facing when using the interface (i.e., the context of use) and to test the design with a sample of users from that population.

Table 1.4 Task Analysis of Various Web Browser Activities with Hypothesized Perception, Cogitation, and Motor Demands

Task	Task or subtask	Perception	Cognition	Motor
1.0	Load the web browser			
1.1	Double click on the web browser icon on the desktop	Visual acuity	Attention, visual search	Coordination, psychomotor speed, fine motor control
1.2	Move the mouse pointer over the "Netscape" icon	Visual acuity	Declarative knowledge, spatial translation	Coordination, psychomotor speed, fine motor control
1.3	In rapid succession, click the left mouse button without moving the pointer			Coordination, psychomotor speed, fine motor control
2.0	Finding information on a page			
2.1	Scan the page for a relevant link	Visual acuity		
2.1.1	In the main browser window, scan the text for information that would lead to the desired information	Visual acuity	Visual search, reading comprehension, declarative knowledge, working memory	
2.2	Click the relevant link			Motor control, psychomotor speed
2.2.1	Move the mouse pointer so it is over the link	Visual acuity	Attention, spatial translation	Fine motor control
2.2.2	Wait until the pointer turns into a pointing hand	Visual acuity	Declarative knowledge	
2.2.3	Click the left mouse button		Declarative knowledge, reading comprehension, working memory	
2.3	Scan the new page for relevant information	Visual acuity	Visual search, attention, working memory	
2.4	Repeat steps 2.1–2.3 until the desired information is found			

1.6.1 Compliance resources

Americans with Disabilities Act Guidelines (ADA)
U.S. Department of Justice

- http://www.ada.gov/

Usability.gov
Bailey, B. (2005). Age-Related Research-Based Usability Guidelines.

- http://www.usability.gov/pubs/112005news.html

Literature Reviews
Arch, A. (2008). Web Accessibility for Older Users: A Literature Review, W3C Working Draft 14 May 2008.

- http://www.w3.org/TR/wai-age-literature/

AARP—Older, Wiser, Wired
Chisnell, D. and Redish, J. (2005). Designing Web Sites for Older Adults: Expert Review of Usability for Older Adults at 50 Web Sites. February 1, 2005.

- http://assets.aarp.org/www.aarp.org_/articles/research/oww/AARP-50Sites.pdf

Ben Schneiderman's recommendations
Leavitt, M. O. and Schneiderman, B. (2006). Research-Based Web Design & Usability Guidelines, U.S. General Services Administration.

- http://www.usability.gov/pdfs/foreword.pdf

A survey of tools to automate initial usability testing of Web displays
Ivory, M. Y. and Hearst, M. A. (2001). The state of the art in automating usability evaluation of user interfaces. *ACM Computing Survey,* 33(4), 470–516.

1.7 Overview of the book

The book is divided into three main sections: a primer on age-related differences in user populations with guidelines and example redesigns, in-depth design sections with examples from several common display genres, and a guide to testing displays with older users. The primer is designed to give more insight into the nature of age-related differences to provide context for the design recommendations. No matter the display, to successfully use it a person must be able to see and hear it (vision, hearing), to be able to think and decide about it (cognition), then to act out on the decision (movement). These fundamental aspects of human behavior are covered in Chapters 2 through 5. In these chapters we discuss the balance of user's capabilities and limitations and how this balance changes with age. Within each chapter examples are provided to illustrate relevance in a display/user interface context.

Specific examples of generic commercial displays illustrate current problems with those displays and potential solutions based on the information in each chapter. The problems are illustrated by interfaces that we have made generic so as not to target any particular company; however, the problems are real. Most are composite examples of many real products that we have encountered. Each of the fundamental ability chapters ends with an applied example of an interface that required that ability. We point out the potential issues older users may have with this display and suggest changes to make it more usable.

It is important to note that age-related differences often interact with each other, and a design that addresses one age difference can exacerbate the problems associated with another difference. For example, the task analysis in Table 1.4 listed the range of abilities required for seemingly simple tasks. Changing a display according to one required ability may adversely affect one of the others. To highlight potential interactions we have included "Potential Interaction" callouts throughout the book. These short sections are meant to draw attention to how a topic we just discussed interacts with other topics in the book. For example, a design change that is good for perception may not be good for cognition, as provided in the first example.

POSSIBLE INTERACTIONS TO CONSIDER

Even a well-meaning change to an interface can do more harm than good. Designers must consider potential unintended consequences that might come from following specific guidelines. The variety of displays and their interface controls dictate that there will almost always be interactions and trade-offs in usability.

The Florida ballots for the 2000 election provided an example of good intentions going awry. The ballot (Figure 1.2) was the result of an untested user design. Knowing Floridians tended to be older, Theresa LePore, Supervisor of Elections for Palm Beach County, Florida, stated that she wanted to increase the font size so the ballot could be easily read. However, just increasing the font size made the ballot so large it could not be used. The "clever" fix was to alternate names on the left and right side, adhering to perceptual needs of older users, but violating their expectations (as well as the expectations of younger users). The creator could claim she followed guidelines to have an acceptable font size, but cannot ignore the ultimate failure of the design.

Think carefully about the consequences of each decision, but also test designs with older users. Throughout this book we provide examples of "Potential Interactions" that go beyond the concept being discussed in one chapter. These boxes will provide illustrations of design variables to consider in a design and highlight possible interactions.

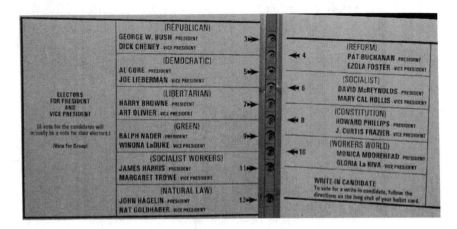

Figure 1.2 The butterfly ballot.

After introducing the fundamentals of age-related change in perception, cognition, and movement, we provide a chapter on performing usability tests with older users. Though bringing in older users for tests is similar in many ways to testing younger users, there are a few differences that we discuss. We also provide advice on recruitment, retaining users for future tests, and simulating their experiences for the designers working on the product.

Our book ends with four chapters that apply the fundamental knowledge from Chapters 2 through 5 and the usability testing information in Chapter 6 to specific products. In these chapters we chose products that could be improved by this process and walk through the usability testing steps from overview of the issues to developing a user profile to performing a task analysis of the system to be improved. We close each of these chapters with a conceptual redesign of the system.

1.7.1 Usability evaluations and redesign chapters

Chapters 7 through 10 provide worked, brief examples of age-sensitive design and evaluation of displays. We called them "conceptual evaluations" for several reasons. First, the actual interfaces being evaluated (feature-phone interface, television set-top box, blood glucose meter, and global positioning [GPS] navigation system) are not specific interfaces *per se* but are generic examples of interfaces. Second, an actual usability evaluation was not conducted (i.e., usability study with older users); instead, we relied on past literature, our own experiences with older adult users, and an expert evaluation (a heuristic evaluation) to determine problems that older users are likely to have, then implemented a redesign that *should* be tested with older users. Each redesign chapter ends with a summary of guidelines about that design that are based on the earlier chapters on vision, audition, movement, and cognition. Last, we realize that there are limitations imposed on the designer or user-experience professional working on interfaces. Software and hardware may be designed by separate parties, and the degrees of freedom to change design aspects may be little to none. However, we hope that the specific targets of evaluation and general comments will be applicable to classes of systems that exist today and in the future so that designers may have maximum impact on products for older adults.

1.7.1.1 Organization of the redesign chapters

Just as a typical usability evaluation might begin with an understanding of the user, each evaluation chapter begins with a user profile that encapsulates the capabilities and limitations of the target user (a typical older adult user of that particular display)—answering the questions of what the user can do, and what he or she wants from the system.

An optional but useful next step after the creation of the user profile is to create a persona or an "archetype" user that encompasses elements of the profile. The amalgam of characteristics is given a name as well as motivations, attitudes, and typical behavior patterns consistent with that user group. Being able to predict how users will react or what they will do next is one beneficial aspect of personas. A user profile is a "summary" of all known data about users, and a persona puts a face on those data. Subsequent usability flaws or possible solutions are then filtered through this "persona." A persona may be useful when answering questions of a subjective nature such as, "Will the user need this function?" "Will the user like this function?" Further information about user profiles and personas can be found in the Suggested Readings section of this chapter and in Chapter 6—Older Adults in the User-Centered Design Process.

Each evaluation also encapsulates scenarios that combine a typical task and context. A scenario represents a compact and easy-to-understand story about the user trying to complete a task and how they react and behave when interacting with a display. The key output of a user profile and scenario will be to answer the question of whether the user is successful in accomplishing their goals, and if not, why. It is best to base use cases on data from current or potential users. We follow each scenario with a step-by-step task analysis of how the display is currently used. For each step in the task, we present an analysis of the potential age-related usability issues of the original design. Then we present the same step redesigned for an older user, both illustrated and with the changes noted in text. Although we based these redesigns on knowledge of usability and aging, the designs were not user-tested for inclusion in this book. Instead, they are provided as illustrative examples of the redesign process for displays targeted to older adults. As with any new design, different prototypes should be tested directly with older users, followed by iteration of the design.

1.7.1.2 Displays chosen for evaluation and redesign

Chapters 7 through 10 cover a range of displays and their accompanying interfaces that are frequently used by older adults. These included a cell phone, a set-top box for cable or satellite television, a blood glucose meter used by persons with diabetes, and an automobile navigation system. All of these systems bring benefits to older users, although in their current state those benefits may not outweigh the usability challenges.

Chapter 7 details the analysis and redesign of a mobile phone with a focus on using the phone to send an e-mail. A phone was chosen because of the utility of the display and because the small screen size brings unique challenges to an age-sensitive design. Because of the screen size, our focus was to improve the perceptual and cognitive qualities of the display. We did this by reducing clutter (especially small icons) and changing the flow of task to take advantage of recognition memory, increased feedback, and close control-to-display mapping.

In Chapter 8, we cover changes to a set-top box display used to record a television program. The redesign is highly cognitive: our emphasis was on improving the feedback and options available to the user. This type of interface is particularly susceptible to increases in poorly integrated "features." A user-centered redesign concentrates on features that match the goals of the user.

In Chapter 9 we cover one of the most commonly prescribed home medical devices to people over age 60. The blood glucose meter, or glucometer, is available in numerous forms and brands. Many of these brands advertise their simplicity and speed of use, but this is often not an accurate reflection of their interface. Further, tasks with the glucometer sometimes need to be performed when a user has diminished cognitive function, as with low or high blood sugar levels. A display such as this is used frequently, such that the user has more of a chance to develop expertise with common functions than for devices such as the set-top box. However, the glucometer can have important tasks, such as calibration, that are infrequently performed and have relatively high demands on cognition. We focused our analysis and redesign on these types of tasks.

Chapter 10 addresses the complexity of a GPS car navigation system. This type of display required careful consideration of the movement involved in configuration as well as the perceptual demands of a small screen situated a fair distance from the user. Though the redesign focused on the task of inputting a destination, the chapter also covers the challenges of designing the auditory information provided by this device.

We know the technology we chose for redesign will change drastically in the future. Some changes will make them easier to use by older adults, and others will make them more difficult. The most important information in these chapters is the *process* of redesign: looking in depth at any product for any purpose in terms of potential difficulties older users might experience. This process requires an integration of general knowledge of age-related change and an understanding of usability techniques and how they may differ for older adults.

1.8 Suggested readings

Craik, F. I. M. and Salthouse, T. A. (2000). *The Handbook of Aging and Cognition* (2nd ed.). Mahwah, NJ: Erlbaum.

Harper, S. and Yesilada, Y. (2008). *Web Accessibility: A Foundation for Research.* Human-Computer Interaction Series, No 1, London: Springer.

Pew Internet and American Life Project. (2009, November 19). 2006 Report on Americans and Their Cell Phones. Retrieved from http://www.pewinternet. org/Reports/2006/Americans-and-their-cell-phones.aspx

chapter two

Vision

We are surrounded by visual displays from the moment we wake up each day. Consider my typical day. When I awoke, the first thing I saw was my digital alarm clock. When I drove to work, I was surrounded by glowing visual displays in my car. The instrument cluster itself is a display of information, but my car also has small liquid crystal monitors to display additional information such as mileage, time, temperature, and average miles per gallon, navigational maps, or radio stations. When I arrived at work, I sat down in front of my dual-monitor setup, which is the most obvious display. However, as I looked around, there were other displays such as my mobile phone and the caller ID display on my office phone. The purpose of each of these displays was to convey information, and each required varying amounts of my attention.

The importance of visual displays cannot be overstated. Although information input and output is multimodal, vision is the most fundamental way in which technology presents information to the user. Some researchers have found that a majority of users' frustrations with technology can be traced back to not being able to see important aspects of the display. There are other ways for a system to provide output other than a visual display, for example the auditory display of my alarm clock or the vibrations from my cell phone. However, visual displays are the most prevalent display type and occur across a wide range of products.

The purpose of this chapter is to highlight important sources of variation in visual abilities that may disproportionately affect older users. The chapter is divided into three sections. First, we provide a general overview of changes in visual abilities that accompany the aging process. A discussion of the functional implications for these changes and how various display technologies impact the user's experience follows. Finally, the principles of good visual display design are illustrated in an example, with some suggestions to mitigate age-related limitations in vision.

2.1   How vision changes with age

Most people are already directly or indirectly familiar with the effect of aging on vision. We may have an older relative or parent who is experiencing difficulties, or we may have experienced them ourselves. A familiar stereotype of an older person is one who wears bifocal glasses. Bifocal glasses contain two levels of correction: one for near and one for far

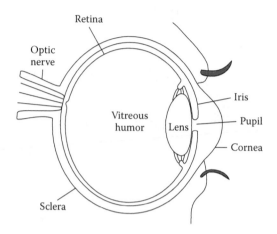

Figure 2.1 Diagram of the eye.

vision. The need to wear bifocals, triggered by the gradual loss in the ability to alter the shape of the intraocular lens for different distances ("presbyopia"), may be the most easily recognizable age-related visual change; however there are other, more subtle changes that affect a user's ability to read a display.

Figure 2.1 provides a short refresher on the anatomy of the eye (Figure 2.1). The human eye is essentially configured as a camera. Light enters a variable-width hole or aperture called the pupil (akin to the shutter of a camera). The size of the pupil opening is controlled by a series of small muscles surrounding the pupil. Light then goes through a transparent flexible lens that can (for those young enough) focus the light onto the back surface of the eye called the retina. The lens is also controlled by a set of small muscles that stretch or push the lens into different shapes. After passing through the lens, the light lands on the light-sensitive retina. From there, light is transformed into signals that get processed further back, in the brain.

Depending on lighting conditions and the observer's mental state, the pupil changes—shrinking in size to let less light enter the eye under bright conditions or expanding to let more light in under dark conditions. However, in the aging eye, pupils have a harder time changing size. Not only do the pupils change size more slowly, they are less able to maximally dilate or open to let in light during dim conditions. The result is that the older eye receives dramatically less light than younger eyes. At night this is functionally equivalent to wearing sunglasses. And the older we get the darker our permanent sunglasses become. Combined with an age-related reduction in pupil size, the retina of a 60-year old may receive as much as two-thirds less light than that of people in their 20s.

Once light enters the eye through the pupil, it passes through the lens. Again depending on conditions (e.g., the distance of the object of interest), the flexible lens changes shape as the eye tries to create a clear focus. This process, called accommodation, happens automatically and outside of conscious awareness. With age, the once flexible lens becomes less able to change shape. The lens is also less able to quickly and effectively change shape to focus the light onto the retina and the result is a nonoptimally focused image on the retina.

In addition, as the lens gets older, it turns from transparent to slightly yellow. This yellowed lens preferentially absorbs blue light, making colors appear less blue and more yellow. The result is that it becomes harder to distinguish between subtle shades of blue. In more severe cases, distinguishing between shades of red and purple become more difficult.

Another common age-related lens condition is cataracts. Cataracts are the gradual clouding of the lens, resulting in a hazy vision that is highly susceptible to glare (reflected light). Although cataracts are highly treatable, many older adults may be in the early stages of cataract development but untreated.

2.1.1   Visual acuity

Acuity is the sharpness with which a person perceives a visual image. It is a measure of the resolving power of vision (the ability to see fine detail). Acuity is measured with the Snellen eye chart (which may be familiar from the optometrist's office) and is expressed in terms of two numbers (e.g., 20/10). Figure 2.2 illustrates the Snellen chart. The observer stands 20 feet away from the chart and reads as far down as possible. The Snellen acuity score represents the farthest that one can read down the chart. The 20/20 level is what a "standard" person (that is, with "normal" acuity) can read from 20 feet away, but healthy, young eyes often exceed this standard level. The denominator gets smaller to indicate better than normal vision. For example, a person with 20/15 vision indicates that what a normal person can see at 15 feet, that person can see from 20 feet away. Similarly, the denominator gets larger to indicate worse than normal acuity. If a person has a score of 20/200, that person can only see at 20 feet what most people can see at 200—quite bad vision! A person whose best optically corrected acuity is worse than 20/200 is considered legally blind in the United States. When acuity is low, it is high-frequency visual information that is lost (e.g., fine lines, hard edges).

A variety of physiological changes in the eyes can lead to a decline in visual acuity with age. Figure 2.3 shows mean visual acuity (y-axis) as a function of age (x-axis). The plot combines the results of many different studies showing that with increasing age, there is a steady decline in visual acuity for most people.

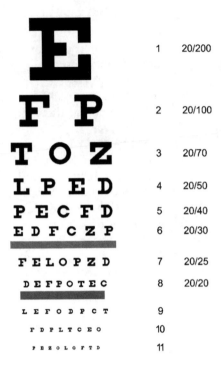

1	20/200
2	20/100
3	20/70
4	20/50
5	20/40
6	20/30
7	20/25
8	20/20
9	
10	
11	

Figure 2.2 Snellen eye chart. (Available on Wikipedia under a Creative Commons Attribution ShareAlike 3.0 license. http://en.wikipedia.org/wiki/File:Snellen_chart.svg. With permission.)

These changes in visual acuity have implications for the use of displays. Studies with older adults using computers have found that in some cases they were not even able to start the experimental task because they could not read the display. More specifically, visual impairment due to age-related acuity loss has effects such as slower visual search time to find icons, and confusion in icon selection. These problems are made worse when the icons are particularly abstract or similar looking (in shape, size, or coloring). To demonstrate this confusion, consider the simulation in Figure 2.4. The figure illustrates that when high-frequency information is removed from icons they become much harder to distinguish from each other.

2.1.2   Contrast sensitivity

Acuity is perhaps the most easily understood and measured aspect of vision. However, vision scientists have found that when it comes to performing everyday tasks such as reading, driving, or using a computer, contrast sensitivity is more important than acuity. Contrast sensitivity is the ability to distinguish between light and dark parts of an image.

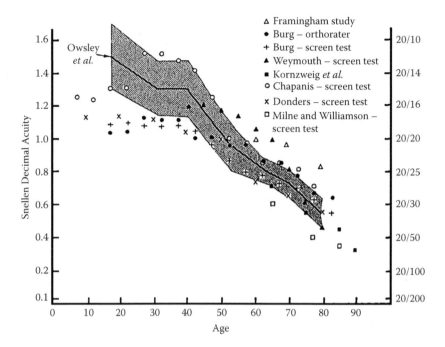

Figure 2.3 Mean visual acuity (*y*-axis; higher is better) as a function of age (*x*-axis). (From Ousley, C. et al., *Vision Research* 23(7), 689–699. 1983. Reprinted with permission.)

Contrast sensitivity is important because many details in everyday life are rarely as high contrast as the Snellen chart. Unfortunately, as with acuity, older adults tend to have reduced contrast sensitivity.

Contrast sensitivity is measured by testing the lowest level of contrast (light and dark) that an observer can distinguish. An observer may be shown letters of varying density and asked to read as far down as possible

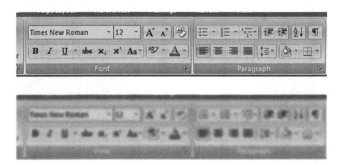

Figure 2.4 Top panel shows an icon palette. The bottom panel shows how the icons appear similar with blur (simulating lowered visual acuity).

QDGFT
CSPQY
LJIFE

Figure 2.5 A simulated contrast sensitivity chart (going from top to bottom, high to low contrast).

(Figure 2.5). Reductions in contrast sensitivity due to age can have pervasive effects on the ability to carry out daily activities (e.g., seeing a dark-clad pedestrian at night). Figure 2.6 shows an example of low-contrast items that might be found on a web page. The upper graphic is a bar graph that illustrates the amount of online storage available, while the lower graphic shows page numbers and current location in a multipaged website. The contrast between the white background and meaningful information (labels, page numbers) is extremely low. This may impede perception, leading the user to devote more attention than is necessary to understanding the graphic. In Figure 2.6 the low contrast is an obvious problem for text readability, and the low contrast shapes also make it difficult for the user to perceive and extract information from graphical elements and understand actionable interface elements.

2.1.3   Visual search

Scanning a grocery store shelf for a specific cereal box is a classic visual search task. Visual search involves moving the eyes and the attentional focus around the scene in search of something specific. The pattern of visual search (how and where to look) depends on the characteristics of the scene and the searcher. A bright red cereal box will draw attention toward

Figure 2.6 Examples of contrast sensitivity issues in displays. The top graphic shows a bar graph indicating storage available in a webmail account. The lower image shows page navigation on a website.

it almost against the will of the searcher. Similarly, looking for a blue box will allow search to be focused only on blue things. Scientists describe two kinds of visual search. The first, exemplified by the red cereal box, is preattentive—described as effortless and not requiring of attention. The second type of visual search is effortful search, the kind of visual search that operates serially and feels difficult. For example, a glowing red neon sign at night is conspicuous and will draw attention to itself. Research shows that there are few age differences in preattentive visual search; that is, both younger and older adults are able to notice and pick out conspicuous elements in the visual environment. However, effortful search shows large age differences, and these differences increase with the difficulty of the search and the number of items to be searched. Designing for preattentive search means to enhance conspicuity, though this is not always possible when many items in a display have equal likelihood of being a targeted choice (such as a list of products on a shopping website) and is usually an effortful search.

In addition to enhancing conspicuity (making the desired object distinct from the surround), another way to enhance preattentive visual search is to utilize the human ability to easily pick out patterns or relationships. Certain configurations of objects have a way of being conspicuous or perceived in a certain way. Early psychologists defined these rules into what are known as Gestalt laws. Gestalt laws are especially useful to know because they are applicable to display design in very straightforward ways. For example, the law of proximity describes that humans perceive objects that are arranged close together as belonging together. Although the left side of Figure 2.7 is perceived as a square made up of dots, the right side is perceived as four rows, each made up of four dots.

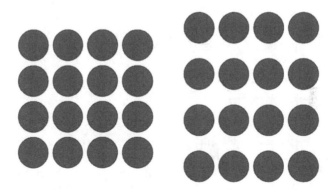

Figure 2.7 The even spacing between dots on the left cluster leads to the perception of a single unit (a square). The smaller horizontal spacing compared to the vertical spacing on the right cluster leads to the perception of four rows instead of a square (or four columns).

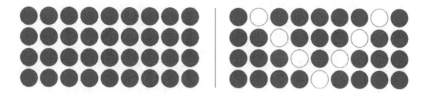

Figure 2.8 Left graphic depicts a solid field of gray dots. Right figure shows how perception of a white "V" is stronger than the background of dots.

A single dot's proximity to its neighbor influences the overall perception of the shape; equal spacing suggests a square, unequal spacing suggests four separate rows.

Similarity is another Gestalt law that influences how we perceive objects. Objects that are similar to one another are perceived as belonging to a unit compared to objects that are dissimilar. For example, the left side of Figure 2.8 illustrates a rectangle made up of gray dots. We perceive this as a singular unit (the rectangle) because of the similarity of the dots to each other. However, if some of those dots are more similar to each other than the background, they are now perceived as a separate unit (the right side of the figure is perceived as a white V embedded in a background of dots).

These laws can be applied in display design to create categories of items so that a user can have all options available in one place, but be able to quickly search higher-level categories before zeroing in on the desired target. For example, on a bookseller's website, many popular books can be shown on a single page. When the icons and descriptions of the books are grouped so that similar books are closer to one another, the user can devote visual search just to the category he or she is interested in. Of course, creating good categories that match the user's expectations of categories is not a simple matter, which we address in Chapter 4—Cognition.

2.2 Interim summary

Physiological changes to the eye related to aging result in less light reaching the retina, yellowing of the lens (making blue a difficult to discern color), and even the beginning stages of cataracts result in blurriness. The eye muscles are also affected; it can be more difficult for older adults to quickly change focus or get used to fast-changing brightness. Some solutions for design include: conspicuity can be enhanced by enhancing contrast and taking advantage of preattentive processes, and effortful visual search can be lessened through application of Gestalt laws.

2.3 Display technologies

User performance and satisfaction with the use of a display depends on the interaction of display capabilities and user limitations. As a complement to better understanding user capabilities and limitations, we must also understand technological capabilities and limitations. Thus it is useful to discuss various display technologies and their general characteristics from an older user's perspective.

A wide variety of display technologies is used in computer and consumer electronics. The three main display technologies in use today are cathode-ray tubes (CRT), liquid crystal displays (LCD), and e-paper technologies. Table 2.1 provides an overview of these three display technologies. The characteristics of these display technologies may interact with age-related changes in visual abilities to make reading more difficult. The underlying technology for display technologies is different, so to facilitate comparisons it is easier to discuss various display-related metrics that apply to each type of display.

Contrast ratio is a ratio, or comparison, of the brightest (white) and darkest (black) parts of a display. The ratio is usually indicated using two numbers: XXX:YYY, where XXX is the brightness level (in arbitrary units) of the brightest part and YYY is the calibrated brightness level of the darkest part. A display with a contrast ratio of 500:1 means that the white is 500 times brighter than the darkest portion of the display (black). Possessing a high contrast ratio (for example, 500:1) indicates a greater perceived contrast. For example, text will stand out from the background more clearly at 500:1 than in display with a low contrast ratio (150:1). A high contrast ratio also helps with the display of subtle shades of gray or color. CRT and e-paper displays have the highest contrast ratio, whereas LCDs can have medium to low contrast ratio. Figure 2.9 illustrates some examples of varying contrast ratio.

Computer displays are made up of millions of individual elements or pixels. Pixel density is a measure of how closely the pixels are arranged on the display. It is indicated in pixels per [square] inch (PPI). CRT and e-paper generally have high pixel densities, whereas LCD pixel density can vary tremendously. Pixel density will affect how sharp a display is perceived to be, as well as its readability (Figure 2.10). A low-density display will appear blocky and jagged, while a higher pixel density display will appear sharper and smoother. Screen resolution is the number of light-emitting picture elements (pixels) in the display independent of pixel density and should not be confused with PPI.

Viewing angle is the maximum angle of the observer relative to the display surface at which a display still appears acceptably clear and bright. The optimal location to view a display is directly in front (Figure 2.11); however, that may not always be possible. The viewing angle measurement

Table 2.1 Summary of Display Technologies and Their Characteristics

	CRT	LCD	E-paper
Typical contrast ratio	HIGH	MED-HIGH	HIGH
Brightness	HIGH	MED-HIGH	LOW
Viewing angle	HIGH	LOW-MED	HIGH
Pixel density (typical)	HIGH	MED-HIGH	MED-HIGH
Diagonal size range	MED	LOW-HIGH	LOW
Typical Resolution	HIGH	MED-HIGH	LOW
Refresh rate	MED-HIGH	HIGH	HIGH for static material (low when display is changing)
Typical application	Computer displays, televisions	Computer displays, televisions, mobile phone screens	Electronic books, mobile phones
Aging Implication	Glare due to brightness may be the primary issue. Anti-aliased text will appear more blurry on CRTs than LCDs. Systems set to a low refresh rate (appearing as waviness or blinking) may cause eye strain during extended reading. Brightness may hinder dark adaption when it is necessary (in vehicles).	Anti-aliased text appears best on LCDs. However, contrast ratio will make some things more difficult to see for older adults who have reduced contrast sensitivity. Brightness may hinder dark adaption when it is necessary (in vehicles).	High contrast ratio and pixel density make readability easy (however, some early e-paper displays have relatively low contrast). Dependence on ambient light makes glare a potential issue.

identifies the point (from either side) at which the visual display will start to degrade from the viewer's perspective. This is often called the viewing "cone." In addition to spatial degradation of the image, other measures of the display degrade with increasing viewing angle (contrast ratio, brightness). CRTs and e-paper have the highest viewing angle (i.e., they can be seen relatively clearly from extreme viewing angles), whereas LCDs have medium to low viewing angle.

Brightness/luminance of a display is measured using the standardized unit of candela per square meter (sometimes abbreviated as CDM2

Figure 2.9 Top left and lower right panels show high contrast ratio; the top right and lower left panels show low contrast ratio.

or cd/m²) or *nit*. The measure indicates how bright a surface will appear to the eye. Display brightness is indicated by their nit values. Higher nits indicate higher perceived brightness at the maximum setting. Typical nit values for computer displays are in the range of 250 to 300 nits, whereas televisions have values of 500 nits or more.

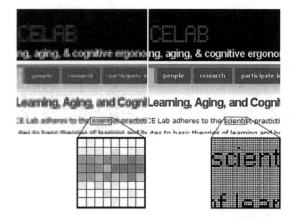

Figure 2.10 Left panel illustrates a display with low pixel density. Note the size of the pixels (squares). The right panel illustrates a higher pixel density (more pixels per unit).

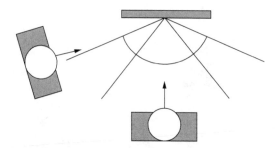

Figure 2.11 Two example viewing cones. Optimal viewing is within the cone (viewing cone depends on the display type).

POSSIBLE INTERACTIONS TO CONSIDER

It might seem that increasing brightness is one way to bypass lowered contrast sensitivity and less light entering the eye. Unfortunately, a display that is too bright can produce other issues for older users, such as glare. Another problem is that older adults are slower to dark-adapt than other age groups. As the light levels in the environment dim, we slowly adapt to the reduced illumination; for example, waking to go to the bathroom at night, we can more easily see the darker surroundings than when quickly moving from a bright room to a dark room. The dark adaptation process occurs more slowly for older adults, and an overly bright display can further slow older adults' ability to dark-adapt when the display brightness changes. A solution can be to avoid fast changes in brightness in a display and provide quickly accessible controls for brightness to allow the user to choose a comfortable level.

With the exception of e-paper displays (which typically do not emit their own light), LCDs and CRTs are light-emitting devices. Reading from such displays for prolonged periods may cause eye strain for some users. In addition, for screens with large diagonal sizes (physical size of the display measured diagonally) or high screen resolutions, a sentence may be much longer than what the reader is accustomed to seeing on paper, making it difficult to track a sentence.

CRTs are the most commonly used display technology for computer displays and televisions. These displays are also called "direct view" because the viewer directly views the lighted image (compared to viewing projected output of an image). CRT displays are distinguished by their glass (sometimes curved) surface, large size, and heavy weight. CRT displays are an old technology, inexpensive, and still very common among home users. They also offer the best contrast, brightness, and color rendition of any display technology. The light-emitting surface also ensures an extremely wide viewing angle.

LCDs are becoming much more common in television and computer display applications as their prices dramatically decrease. Their primary characteristic is that they are lightweight, which allows them to be used in many more applications than CRTs such as portable devices and in-vehicle displays. However, because the lighting source for LCDs is behind the actual display, the image may quickly degrade unless the user is viewing from directly in front of the display. The contrast ratio also varies depending on the specific display. LCDs are available in the widest range of sizes from small mobile phone display (sub-inch) to home theater televisions (several feet).

E-paper displays are the newest type of displays and therefore the most expensive on a dollar-per-diagonal-size basis. The distinguishing characteristic with e-paper displays is that they have an extremely high pixel density compared to CRT and LCDs and a contrast ratio which can, in some cases, replicate the look of printed type. This makes them ideal for applications that primarily involve reading, such as electronic book reading devices. Another characteristic of e-paper displays is that they typically do not include their own light source and instead use reflected light from the environment.

2.4   In practice: presenting information on the web

This section will discuss the presentation of information on the web as an example that brings together the issues discussed at the person-level and how they interact with system-related issues. Our goal was not to examine presentation of the web per se, but to use it as an example that integrates broad ideas applicable to many different types of displays.

2.4.1   Presentation of type

Increasing numbers of older adults use the internet and may rely on it for communication with family and friends and some activities of daily living. Thus it is critical to understand how the design of websites might interact with age-related changes in visual abilities. As one example, the simple act of reading the contents of a website may be affected by some of the factors that have been discussed in this chapter.

The size of the type has the largest effect on user performance in reading on the web. In general, larger-sized type is easier to read than a smaller size. In addition, user studies suggest that sans serif type is perceived to be more legible than serif fonts. An illustration of serif and non-serif (sans serif) fonts is shown in Figure 2.12.

The presentation of text on the web, unfortunately, is something that is affected by many factors besides type size. First consider the different operating systems that people may be using. The presentation of type is affected by individual settings on a user's computer. For example, as a default, the Macintosh and Windows Vista operating systems anti-alias all text. Anti-aliasing provides a smoother look to the text by incorporating

Serif
Sans Serif

Figure 2.12 Top of panel shows a serifed font. Note the extra strokes at the end of the letters. Lower portion of panel shows a sans serif font without any extra strokes on the letters.

pixels between black and white ones or using a complex dithering scheme (Figure 2.13), shown to improve reading speed for younger adults. However, although anti-aliasing can make the edges of letter forms look smooth, it may cause smaller text to appear smudged and blurry. Newer anti-aliasing techniques such as subpixel font rendering (e.g., Microsoft's Cleartype) alleviate some of the small text blurriness, but only work on certain types of monitors and may still add to the perception of blurriness for some users.

Unfortunately anti-aliasing is a complicated issue because anti-aliased text appears different depending on the specific computer configuration and settings. Consider the panels of Figure 2.14. The left image is how a web page is rendered in the Internet Explorer or Firefox web browser. However, that same web page in Apple's Safari Browser appears denser, darker, and blurrier in the right panel. The size of the individual letters also appears smaller and more tightly spaced in Safari than in Internet Explorer. This presentation may be preferred for some people, but the designer should be aware of presentation differences due to the viewing equipment. The Suggested Readings section contains links to resources that let the designer preview how certain combinations of fonts and weight will appear on different browsers and operating systems.

Anti-aliasing technologies applied to type were originally created as a solution to type that appeared jagged and rough, not necessarily as a way to increase user performance. This is a potentially sensitive issue as small blurry text may impact older adults more than other age groups. Little research exists examining reading and comprehension of anti-aliased text

Figure 2.13 Magnification of anti-aliased (subpixel rendering technology) text in an application and on a web page. Note the intermediate grayscale pixels on the edges of letter forms.

Figure 2.14 Text anti-aliased by Internet Explorer or Firefox shown on left, Apple Safari on right. Note the denser text, and letter height of the paragraph text compared to the left side.

in older adults. What is known is that with increasing age, sensitivity to blur increases. For older adults, while they may still perform adequately in blurry text conditions, they may prefer it less than non-anti-aliased text.

Finally, another layer of complexity is that text on the web is sometimes presented in the form of a graphic instead of text. In those cases, designers may have applied anti-aliasing and then locked it in the form of a graphic. This prevents the user from applying custom specifications to this text.

The designer has a great deal of customizability with the display of text. Table 2.2 lists some of the parameters that can be adjusted, an example, and any aging implications.

2.4.2 *Organizing information on a page*

The web is a mature technology and has developed particular conventions that users expect to be followed. For example, clickable links should be visually offset from the main text though spacing and coloring. Blue-colored text used to be the *de facto* standard to indicate hyperlinks, but this is increasingly less common. Almost anything is clickable—whether it is blue or not—it is up to the user to discover what is clickable.

Table 2.2 Various Font Properties

Property	Example	Aging Recommendation
Font family	Serif "You are old, Father William," the you "And your hair has become very white And yet you incessantly stand on your Do you think, at your age, it is right?" Sans serif "You are old, Father William," the yo "And your hair has become very whi And yet you incessantly stand on yo Do you think, at your age, it is right?	Sans serif fonts have been shown to be slightly preferred over serif fonts by older users. This preference may be due to the reduction in visual clutter and complexity induced by the font strokes in serifed fonts.
Font variant	Normal "You are old, Father William," the you "And your hair has become very white And yet you incessantly stand on your Do you think, at your age, it is right?" Small caps "YOU ARE OLD, FATHER WILLIAM," THE YOU "AND YOUR HAIR HAS BECOME VERY WHITE; AND YET YOU INCESSANTLY STAND ON YOUR DO YOU THINK, AT YOUR AGE, IT IS RIGHT?"	Reading all uppercase or small caps is more difficult than reading regular case. This may be due to the similarity in all cap letters (high similarity requires more attention to process).
Line height (line spacing)	80% "You are old, Father William," the young "And your hair has become very white; And yet you incessantly stand on your hea Do you think, at your age, it is right?" 100% "You are old, Father William," the you "And your hair has become very white And yet you incessantly stand on your Do you think, at your age, it is right?" 200% "You are old, Father William," the young man "And your hair has become very white;	Suboptimal line height (too much or too little) has been shown to negatively affect perceptions of readability as well as eye strain. The effect is likely exacerbated with aging. Further work needs to examine what is optimal as it is likely to change depending on the task.
Letter spacing (kerning)	Normal "You are old, Father William," the young "And your hair has become very white; And yet you incessantly stand on your he Do you think, at your age, it is right?" 0.1px "You are old, Father William," the you "And your hair has become very white And yet you incessantly stand on your Do you think, at your age, it is right?" 0.3 px "You are old, Father William," the you "And your hair has become very white And yet you incessantly stand on your Do you think, at your age, it is right?"	Like line height, reductions in letter spacing create increased density which requires too much attention to read for older adults. Too much letter spacing results in difficulty because words are no longer perceived as units. Extra attention is required to merge the individual letters into words.

Source: Adapted from Chaparro, B., Baker, J. R., Shaikh, A. D., Hull, S., and Brady L. (2004). Reading Online Text: A Comparison of Four White Space Layouts. http://psychology.wichita.edu/surl/usabilitynews/62/whitespace.htm.

While it may be debatable as to the appropriateness of certain web conventions, it is clear that they do exist. In many computer-based training programs, new users may still be learning these conventions and so expect to see them. Table 2.3 shows some existing web conventions that may interact with visual capabilities and their applicability to age-sensitive design.

It is more difficult to learn, particularly for older adults, how to use an interface when each website differs in conventions and organization (further discussion can be found in Chapter 4). This is not to suggest that there is no room for creativity in the design of websites. In fact, with an increasing advancement in cascading style sheets (CSS) and awareness of the benefits of separating content from formatting and design, it is now easier to create customized sites with little effort. A simple example of this is the relatively recent feature found on some websites of giving users control over font sizes. Although this was always possible at the browser level, changing font sizes using the browser sometimes caused unwanted effects such as ruining the overall layout. Now font size can be changed when needed without breaking the layout of the website.

In addition to different font size preferences, users will be running at different screen resolutions, with the most common now being 1024 pixels wide by 768 pixels tall. Even so, many users do not run their browsers at full screen (maximized). Informal observation shows that older users tend to keep their browsers windowed at a size that is smaller than the maximum screen size. They may do this because they are not aware of the capability to maximize or they may do this more proactively as a way to keep more information within easy view. One way to accommodate this behavior is to program websites with fluid rather than fixed layouts. A fixed layout is a website specifically designed for a specific resolution that does not change based on the size of the web browser window. Alternatives are liquid layout web pages (Figure 2.15). Liquid layouts adapt themselves to the width of the browser; expanding text areas and reflowing text. This is beneficial for older adults or other users who may not utilize the full screen resolution. It prevents parts of the website from being "below the fold" or obscured by a small viewing window.

One method of liquid layout is to create a website that automatically reformats text so it is most easily read and understood through a screen reader. Changing visual information to audio is useful for persons unable to read the screen themselves, but it is not enough to have the machine read the words to the user. It is important that the information appears in an order that makes sense, hence the need for careful reorganization of the site for a screen reader. We discuss the challenges of organizing website information and providing interaction techniques in Chapter 4, and

Table 2.3 Examples of Commonly Used Web Conventions and Their Aging Implications

Convention	Example*	Aging Implication
Blue: hyperlinks Purple: visited links	**Settings	My Account	Help	Sign out** **World »** · Afghanistan tops NATO summit agenda · Gang of villagers chase away Google car **Health »** · Two new reasons to worry about insomnia · iReporter sets out to connect with Korean roots **2 million jobs lost in 2009** Job losses continued to mount in March and unemployment hit a 25-year high, according to the government's latest reading on the battered labor market. For the first three months of the year, 2 million jobs have been lost, and 5.1 million jobs have been lost since the start of	This is one of the oldest web conventions. However, colors in the blue spectrum are more difficult to see and distinguish from other shades of blue. This is problematic when subtle shades of dark blue are used. Underlining is also conventional, but it makes text less legible for older adults by increasing visual clutter.

* Color not represented here. Titles are colored blue.

Hierarchical text structure

Contents [hide]

The Cycle of Human Factors

Human Factors involves the study of factors and development of tools three goals of human factors are accomplished through several proced (brain and body) and the system with which he or she is interacting. A in the human-system interaction of an existing system. After defining order to implement the solution. There are as follows: •*Equipment Des* •*Task Design*: focuses more on changing what operators do than on c other workers or to automated components. •*Environmental Design*: ir noise in the physical environment where the task is carried out. •*Train* will encounter in the job environment by teaching and practicing the ne recognizes the individual differences across humans in every physical performance can be optimized by selecting operators who possess th

Human Factors Science

Human factors are sets of human-specific physical, cognitive, or socia

Web pages with an extensive amount of text often present it hierarchically. This hierarchy should be as visible as possible through the use of table of contents or clearly demarcated headers. Deep hierarchies should be avoided because they will be difficult to see all at once. Whether organized text is helpful or not depends on the level of knowledge the intended user has about the topic.

(*continued*)

Table 2.3 Examples of Commonly Used Web Conventions and Their Aging Implications (Continued)

Convention	Example	Aging Implication
Buttons		Button presentation varies wildly by website and even within websites. At a minimum, buttons within a website should be as consistent as possible. Consistency is a cardinal rule in user interface design and is even more important when designing for older users.
Placement of primary (or global) navigation		It has become customary, at least in cultures that read from left to right, to place navigation on the upper or left-hand section of websites. In complex sites with many pages, it would be wise to stick with this convention, because users may expect to find it there.

Secondary (or local)
navigation

On the shopping website
Amazon, the primary
navigation is presented on
the left (a fly-out menu).
Hovering over an option
brings a second-level
navigation that flies out
(movies, TV, etc.).
However, too many nested
options can be confusing (see
Chapter 4), and the drop-
down or fly-out menus may
not be obvious at a glance.
Fly-out menus depend on the
user holding the pointer over
the main option, something
that requires relatively fine
motor control (see Chapter 5).

(continued)

Table 2.3 Examples of Commonly Used Web Conventions and Their Aging Implications (Continued)

Convention	Example	Aging Implication
Alternate page-level navigation	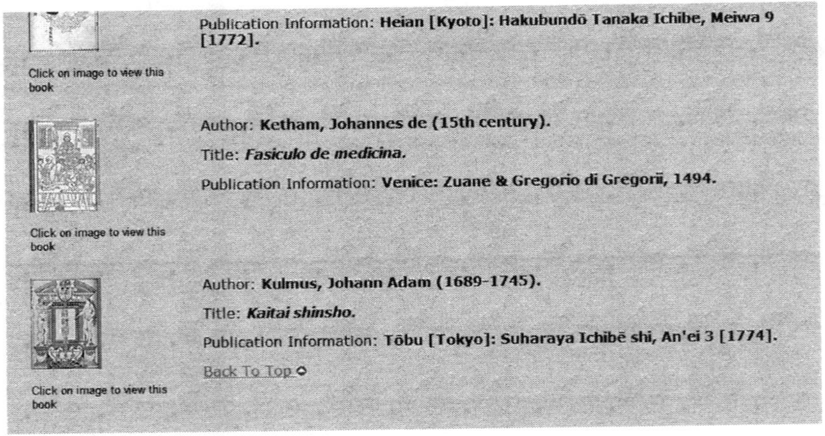 Click on image to view this book Publication Information: Heian [Kyoto]: Hakubundō Tanaka Ichibe, Meiwa 9 [1772]. Click on image to view this book Author: Ketham, Johannes de (15th century). Title: Fasiculo de medicina. Publication Information: Venice: Zuane & Gregorio di Gregorii, 1494. Click on image to view this book Author: Kulmus, Johann Adam (1689-1745). Title: Kaitai shinsho. Publication Information: Tōbu [Tokyo]: Suharaya Ichibē shi, An'ei 3 [1774]. Back To Top	A technique of including a shortcut to return to the top of a long document. For very long documents this may be especially beneficial for older adults because the alternative is to navigate to the slider with the mouse, which may be a difficult task for older adults. Clicking a link is always easier than scrolling with the slider.

Prominent search
box

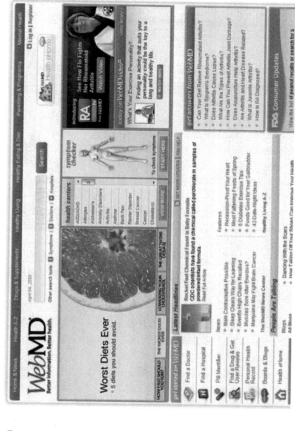

If the site contains extensive
written information, a search
function should be provided,
but it should be located
prominently.

There are other issues with
relying on search as the
primary information-retrieval
mechanism, which are
discussed in Chapter 4.

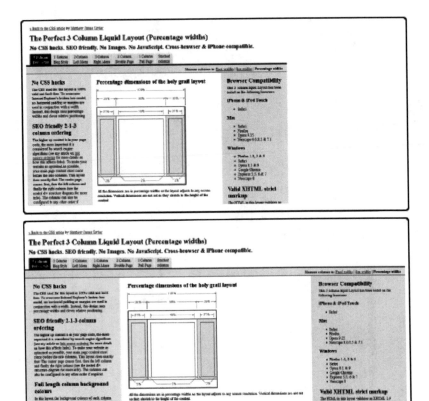

Figure 2.15 Top: A liquid layout at a narrow browser width irrespective of the browser window size. Bottom: A liquid layout with a wider browser width. Note that each section has proportionally expanded.

the importance of creating a low-demand interface is only more important for those using a screen reader.

2.5   General design guidelines

This chapter focused on the effect of vision, and how the visual aspects of the web can interact with aging to produce difficulties. Further discussion of more fundamental aspects of the web (organization of information) are discussed in Chapter 4.

- Background images should be used sparingly if at all because they create visual clutter in displays.
- High contrast should be maintained between important text or controls and the background.

- Older users vary greatly in their perceptual capabilities; thus interfaces should convey information through multiple modalities (vision, hearing, touch) and even within modalities (color, organization, size, volume, texture).
- If the audience is expected to include older users (or is targeted toward older users), follow established web conventions so that users can use their prior experience or training.
  - Within a website, consistency should be the highest priority in terms of button appearance and positioning, spatial layout, and interaction behavior.
- The purpose of a website is to give users what they came for, not make a game of finding it. Reduce the user's need to hunt for information. Older users are likely to have a reduced tolerance for discovery and quit instead of hunting.

Information should be presented in small, screen-sized chunks so that the page does not require extensive scrolling. If this cannot be helped, alternate ways of navigating (such as table of contents) or persistent navigation that follows the user as they scroll should be provided.

2.6   Suggested readings

Adobe Systems (2010). Adobe BrowserLab. (Adobe BrowserLab allows designers to see exactly how their web pages will look in a wide variety of web browser and operating system combinations. In testing and free as of January 2010.) http://browserlab.adobe.com

Browsershot.org has similar functions as Adobe BrowserLab. http://browsershots.org.

Coarna, G. Readable App. (A web-based tool to enhance the readability of text-heavy websites.) http://readable-app.appspot.com/

Jacko, J. A. and Leonard, K. V. (2006). Satisfying divergent needs in user-centered computing: Accounting for varied levels of visual function. In R. C. Williges (Ed.), *Reviews of Human Factors and Ergonomics* (pp. 141–164). Santa Monica, CA: Human Factors and Ergonomics Society.

Perez, A. M. (2008). Common fonts to all versions of Windows & Mac equivalents (A web-based tool to visualize how fonts will appear on Mac and Windows operating systems). http://www.ampsoft.net/webdesign-l/WindowsMacFonts.html

Ousley, C., Sakuler, R., Siemsen, D. (1983). Contrast sensitivity throughout adulthood. *Vision Research* 23(7), 689–699.

chapter three

Hearing

Hearing is an important sense for the use of many displays. For example, last week I needed to set an unfamiliar alarm clock in a hotel room. I needed hearing acuity in connection with vision to determine whether I sufficiently pressed the buttons to change the time, because they responded with electronic beeps but little physical movement. When setting the alarm level, it was my ear that let me know when to stop turning the volume knob. Using the alarm's radio setting required a sensitive ear to tune in a channel via an analog dial.

Hearing is an often-overlooked sense when compared to vision, but this is most likely because the ear can be a more subtle aid when using a display. In the example of the alarm clock, the sound of a button click provided feedback, but there was also visual feedback of the number changing. It is temping to think that hearing is extraneous, but a missing auditory component changes interaction with a system. Consider the use of a "silent" system, such as a keyboard designed to muffle key clicks. Users often report frustration with these systems and use more force than is necessary to activate the keys. Our anecdotal observation of wait staff in restaurants using large touch screens is that they tap the screen with enough speed and pressure to produce a sound in what appears to be an attempt to ensure activation, although sound is not required to activate the display. Hearing is a sense fundamental to every-day function, as noted by Helen Keller: "Loss of vision means losing contact with things, but loss of hearing means losing contact with people." With modern technology, losing hearing now means losing contact with people *and* things. Hearing is often seen as a backup sense to vision, though it has advantages over sight; it functions in 360 degrees around the body and offers higher acuity for determining fast or small changes in a display.

The purpose of this chapter is to provide an overview of the auditory sense and highlight how this sense can change with age or exposure to loud noises. The link between this general information and display design is made explicit throughout the chapter and ends with an integrated example of designing for age-related hearing loss.

3.1   How hearing changes with age

In this chapter we will provide a short primer concerning the ear and what hearing-related changes can occur as people undergo normal aging. When it comes to age-related hearing loss, the most important part of the ear is the cochlea, which looks like a spiral shell. Inside this "shell" are fine hairs, called cilia or hair cells (Figure 3.1). Sound waves vibrate fluid inside the cochlea that vibrates the hairs, signaling to the brain that a particular pitch has been played. The hairs in the widest part of the shell detect the lowest sounds, while the hairs in the narrowest part of the shell detect the highest pitches.

Even the most perfect, undamaged ears can only detect a small range of possible sounds—from about 20 to 20,000 Hz. Human speech occurs almost exclusively from 300 Hz to 3500 Hz, with most speech below 1000 Hz. With increasing age, the hair cells suffer damage from exposure to loud or nonstop noises, and unfortunately, cannot repair themselves. Hair cells can also be damaged by drugs, such as over-the-counter pain medication and some antibiotics. The important point is that a small range within the cochlea can be damaged by loud or repeated exposure to a particular pitch. This creates a "hole" in the range of sounds people can detect and is one reason some speech is more difficult for older adults to process.

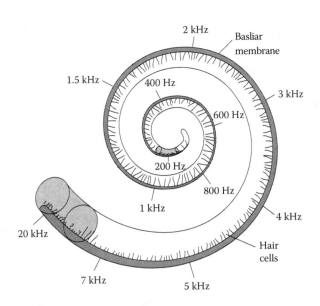

Figure 3.1 General structure of the human cochlea, showing damaged cilia for low- and high-frequency sounds (dark grey areas).

One of the first symptoms of advancing age is a loss in hearing. However, when building an interface with an auditory component, louder sound is not necessarily the answer. There are certain ranges of sound where loss is more common in old age. In the most general sense, older adults can have difficulty hearing the extremes of sound—both high pitches and very low pitches. High-pitched noises, commonly described as a "whine" or "shrill," are generally the first to be lost. Making an interface louder may ensure all components can be heard, but some sounds may become audible while others become distractingly loud.

3.1.1 Pitch perception

How the human brain processes sound, once it is detected by the ear, enters the realm of *perception*. Generally, sensation refers to the actual stimuli that enter the human sense system. For example, the frequency of the sound can be represented by a number, analyzed, doubled, and so on. The perception of that sound is on a different scale from the sensation. For example, based purely on the input, a 2000 Hz tone should be twice as high as a 1000 Hz tone. However, when people are asked to rate those tones, they do not perceive the pitch of the tone to be on a linear scale and do not perceive the frequency as having doubled. Instead, they follow a power law function, where it takes a much larger than double increase to produce the perception of "twice as high."

Another example of where perception is more important than pure sensation is "loudness." Loudness is a subjective measure of sound and is different from measures of sound pressure level, which is measured in decibels (dB). Unfortunately, because loudness is a subjective measure, there are no certain ways of determining how loud a display will sound to any particular person. There are ways of transforming physically measurable attributes of sound to perceptual loudness units, which allow a designer to create adequately loud displays rather than relying on the subjective assessment of what seems "loud enough." Figure 3.2 shows equal loudness curves by decibel sound pressure level (dB SPL). Decibel sound pressure level can be measured with a sound pressure meter in any auditory display and is shown on the y-axis, ranging from below 0 dB to about 130 dB. However, sound pressure is not the only contributor to loudness. Also important is the frequency of the sound, measured in hertz (Hz), which represents the number of cycles per second. Frequency is listed on the x-axis. The lowest frequencies shown are about 20 Hz, and go all the way up to 20 kHz. Clearly, the x-axis is not a linear scale. The human ear is most sensitive at the lowest point on each curve, which maps to about 3500 Hz.

Loudness can be represented by standardized perceptual units called "phons," and the curves in Figure 3.2 represent the sound pressure level

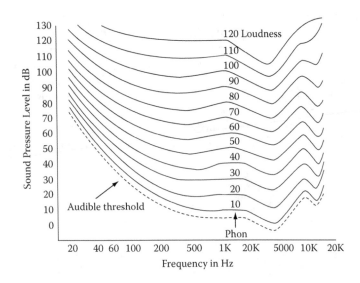

Figure 3.2 Sound pressure level chart showing how frequency and dB combine to form phons, or indicators of "loudness." These are commonly referred to as "equal loudness curves" and "Fletcher-Munson curves." (Adapted from Fletcher, H. and Munson, W. A., *Journal of the Acoustical Society of America*, 5, 82–108, 1933.)

and frequency where a phon is heard. One phon is equal to 1 dB SPL at a frequency of 1000 Hz on a given curve. It is necessary to map loudness contours because the human ear is not equally sensitive to all frequencies of sound. However, this mapping has only been done for healthy young ears, typically between age 18 and 30. This type of chart would look different for someone with hearing loss.

3.1.2   Loudness

Loudness is a surprisingly complex acoustical phenomenon. We present some simple ways to determine the actual and prescriptive loudness levels for an interface and a heuristic for altering loudness, as with a volume knob. Loudness is typically illustrated by example: 35 dB represents a whispered voice from a couple meters distant, while 60 dB represents a typical conversation level, and 100 dB represents the sound of a jackhammer. One of the first steps in creating an auditory display is to measure the loudness of the environment using a sound level meter to ensure that the display volume is above this ambient noise level. Also, once a display is created the loudness of the display can be measured. Basic sound level meters are inexpensive and can be purchased at most electronics stores. On the meter there are several mode settings. Choose the "A" weighting to approximate the frequency weighting of the human ear. Although there are many charts available that give representative examples of loudness levels for different sound sources, the only way to be sure of the level of background noise or the level of display audio is by using a sound level meter.

As mentioned previously, decibels and frequency interact to form the subjective perception of loudness. Whether a sound is loud enough to damage the ear can be measured in decibels alone, and we provide resources for making certain interface sound levels will not cause damage. This is especially important for any auditory display that uses headphones or speakers that are physically close to the ear because it is easy to deliver sound outside safe levels when using headphones in a display. The first rule of thumb is that 90 dB is the upper limit for safety, especially when using headphones or if the user will be exposed to that sound level for long periods of time. The Occupational Safety and Health Administration of the United States released a table that helps in determining safe sound levels according to the length of exposure (Table 3.1). In addition to these duration limits, it is never appropriate to have an alarm over 140 dB, because even one second of exposure at that level can cause damage.

It is not uncommon for a person with age-related hearing loss to require a sound level over 90 dB (especially when there is background noise), so limiting audio to below the 90 dB limit will exclude some users. A rule of thumb is that 85 dB is close to the level suitable for many older users, and 70 dB is suitable for most younger users. However, if the audio is received via headphones and is at levels above 90 dB, the designer runs the risk of harming the hearing of users who have no hearing loss. The best option is often to have the audio levels controlled by the user via a visible and easily manipulated control. This audio should reset to below 90 dB between users to keep the settings chosen by one user from harming the next.

Table 3.1 Safe Sound Levels by Exposure Length

Hours of Permitted Exposure	Sound Level dBA
8	90
6	92
4	95
3	97
2	100
1.5	102
1	105
0.5	110
0.25	115

Source: OSHA Standard 1910, 95(b) (2).

Because loudness is a perceptual quality it can be difficult to map a sound onto a display control so that it increases in volume as the user expects. For example, if the intensity of a sound is increased by a linear (equal-interval) increment with each click of a volume knob or volume slider, the user will perceive little change in loudness until reaching the upper limits of the control. At this point loudness will increase dramatically with each increment. For this reason, volume controls need to be designed to accommodate perceptual changes in loudness to make them appear linear to the user. This can be done via a logarithmic transform to the intensity levels. This reduces the change in intensity at the top of the scale and provides a perception of approximately equal-interval loudness increase with each increment.

There are many available equations for intensity transformations, some specific to certain degrees of hearing loss. These degrees are slight, mild, moderate, moderately severe, severe, and profound. It is possible to guess what the older ear would require differently based on the changes. There are several models that can be used to predict loudness for those with some hearing loss. Papers with the appropriate formulas are cited in the Suggested Readings for this chapter and marked with an *. Many hearing aids use these formulas to determine how "loud" to make the sounds they transmit from air to ear.

3.1.3   Sound localization

Determining the location of a sound in horizontal space requires both ears. Our brains are able to analyze the difference in volume perceived by each ear and orient to the sound. This is why it is much easier to localize sound that is to the left or right; sounds directly in front or behind a listener can be confused with each other. The location of a sound in vertical space is cued by the outer ear and often occurs above 5 kHz, a common point where age-related changes in hearing ability are common. This ability is altered by hearing loss in one ear or differential loss in both ears. It is also negated when using a telephone: when a sound is artificially processed, such as through a cell phone, additional frequencies are lost or eliminated.

Even when older users do not have any documentable hearing loss, it is still more difficult to understand certain sounds in speech due to age-related changing of shape in the inner ear. This phenomenon is known as *phonemic regression*, and the practical implications are that speech really does need to be clear in terms of reduced background noise for older users.

3.1.4   Sound compression

A large majority of output sound from technology (e.g., computers, telephones) is digitally compressed in some way. The need to save bandwidth when playing sound remotely is a driving force for why audio is often compressed before transmission, then uncompressed for the listener. In essence, compression algorithms remove portions of a sound that are "unnecessary" for our perception of that sound. Because the human ear is so limited, a lot of extra information can be removed before the ear detects differences. There are two examples of common compression that illustrate how older adults have been overlooked in the building of compression algorithms: Mp3/Mp4/WMA files and cell phones.

3.1.5   Mp3s, cell phones, and other compressed audio

Moving Picture Experts Group Layer-3 Audio files (Mp3s) are one example of algorithmically reducing the frequencies in audio, typically to make audio storage smaller and less expensive. Most of the design principles for Mp3s can also apply to cell phones with their reduced frequencies and bandwidth. As discussed earlier, the human ear only uses a small range of the available frequencies of sound. What humans can hear is termed the *auditory resolution*. By using psychoacoustic methods, compression algorithms can group similar frequencies and dispose any that would have been masked by other frequencies. However, as audiophiles attest, digitizing sound and removing frequencies does change the quality of sound. Also, Mp3s and other compression algorithms are not equal; they depend on the sampling rate. When choosing a compression algorithm, it is important to research the algorithm for the audio display, because they all have pros and cons and are always being updated.

3.1.6   Background noise

In the kinds of sound used for alerts, such as noticing mechanical problems with autos or computers, there is a general rule that any alarm should be 10 dB above the background noise. Of course, background noise can range from 5 dB in a library to 90 dB on a construction site. This is one reason why it is so important to understand the environment where users will interact with a display. Will it be an ATM machine on a busy street or a quiet one? Indoors or out? Will the user have ultimate control over the loudness? For example, a sound on a website is restricted by the volume settings for the user's computer speakers. In one research study by Berkowitz and Casali, older users were not able to hear a ringing phone with an electronic beep as the notification when there was background noise present.

Added to this general rule, older adults have more difficulty inhibiting background sounds. Like their problems with synthesized speech, this is linked to an age-related cognitive decline: attentional control (Chapter 4.) Someone with high attentional control can choose to attend to one stimulus out of many, for example, a single conversation in a crowded room. As attentional control declines, so does the ability to inhibit all of those other conversations and background noise. This can overwhelm the attentional resources of the listener and contribute to poor comprehension.

3.2   Interim summary

In summary, changes in hearing often accompany aging. Designing around these changes requires some knowledge of physiology as well as some technical expertise in frequency and loudness calculations. The addition of other noise, whether in the background or in the display, as well as many techniques used to compress sound for delivery through technology, can adversely affect those with hearing loss more than those without. The next section discusses some of the technologies developed to assist persons with hearing loss and their effect on design.

3.3   Accessibility aids

Accessibility aids for those who have experienced hearing loss can come from many sources. Some of the sources, such as hearing aids, boost the signal to the auditory system. Others work around hearing by providing other avenues of interaction, such as visual phone systems. Each type of aid is defined then discussed in terms of the interaction with display design.

3.3.1   Hearing aids

People tend to put more significance into vision loss, but hearing loss is an isolating and life-changing condition. Older adults with hearing loss tend to have more medical problems in general, even when those problems are not directly related to hearing. For example, the Centers for Disease Control and Prevention report that older adults with hearing loss have more trouble engaging in activities such as walking and getting outside of the house on a daily basis. It is likely that hearing loss inhibits a number of daily activities that can be undertaken by a nonimpaired older adult.

Roughly 40% of older adults have some reported hearing loss, but only 40% of those use hearing aids. The following is an overview of how a hearing aid works. This is necessary to understand how a hearing aid can conflict with certain auditory display designs.

All hearing aids amplify sound, increasing the perception of volume. Because hearing loss often occurs when the tiny hairs that detect sound via movement die, this increase in volume increases the chance that any remaining hairs will respond to the sound at that frequency. If all the hairs were truly gone for a frequency, the hearing aid would not help.

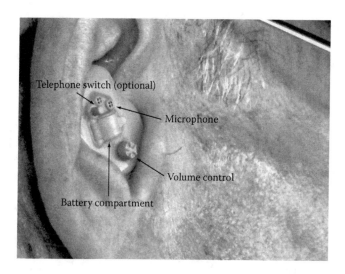

Telephone switch (optional)

Microphone

Volume control

Battery compartment

Figure 3.3 Typical molded hearing aid with controls. (Available on Flickr under a Creative Commons Attribution-No Derivative Works 3.0 Unported. (http://www.flickr.com/photos/portland_mike/2993507037/) With permission.)

Figure 3.3 illustrates a common type of hearing aid used by older adults, the in-ear aid. The aid is molded to fit the ear of the individual. Some aids fit completely in the ear canal and cannot be seen, while others have a battery pack that fits over the top and back of the ear, making them fairly large devices. This range of hearing aids must be considered when an auditory display, such as one presented through a phone, must be in close proximity to the aid. The device in Figure 3.3 exemplifies most of the attributes of an aid that should be considered during the design process. The device in Figure 3.3 is an in-ear aid with several controls available. The volume is the most frequently used control on the device and is operated by turning. The door to the battery pack may be flipped open. The microphone picks up sound to send to the amplifier inside the aid and then into the ear. Some aids have a "telephone mode" that eliminates all ambient noise and allows the focal sounds from the phone to come through.

With larger hearing aids older adults may have trouble using phone-based interfaces. Having a phone touch the ear does not affect hearing much, but having a phone touch the microphone on a hearing aid creates noise. However, holding the receiver further from the ear quickly diminishes the sound and clarity of the phone signal. Being aware that some users may exhibit this behavior can encourage the correct user-testing scenarios.

Another problem for aid wearers is a high, piercing sound known as "feedback." Feedback occurs when sound from the speaker (in this case, the hearing aid speaker) is detected and fed back through the hearing aid microphone. This amplifies the sound into a painful, high-pitched whine.

Feedback is more likely to occur when users need to turn up their hearing aids to maximum volume to detect quiet sounds.

One last thing to consider about hearing aid-wearing users is that an aid does not "cure" the hearing problem. Sounds literally do not "sound" the same when filtered through an aid as they do when heard by a healthy ear. The wearer's own voice may sound different through the aid; while a hearing healthy person can ignore his or her own voice, the hearing aid wearer can detect that voice more like the voice of another person. It is possible to experience partially what hearing is like for an aid-wearer through a normal set of earplugs. The earplugs mimic the hearing-aid effect of "occlusion." Occlusion makes the earplug wearer's voice sound different and other low noises (around or below 500 Hz) will have more of a "booming" effect than without earplugs. This effect can be measured as increasing sound pressure at that level by about 20 dB. As technology progresses this may become less of an issue but should still be considered during design. For hearing aid users, it is important to design displays and interfaces that do not aggravate the side effects of hearing aids and instead work with the aid to produce the most comprehensible sound possible.

3.3.2   Telephony services

TTY stands for teletypewriter and is a service used by many people with hearing loss. Generally, the hearing-impaired individual has a TTY phone with a screen and calls the service. The operator on the other end of the line will connect through to the number being called, and then type in whatever the other party says. This text appears on the screen of the phone. Essentially, the TTY operator is a speech-to-text translator. Other terms for TTY are "text telephone" and TDD. TRS, or Text Relay Service, involves an operator who acts as a speech-to-text intermediary.

Automated menus need to accommodate TTY phones and services. For example, imagine calling a company with a telephone menu system. The menu is spoken, and the TRS operator must type the options for the hearing-disabled caller. The caller must read the options, make a choice, and possibly relay that choice to the TRS operator, who inputs the choice. If the timeout for a menu is short and the call is disconnected (e.g., "I'm sorry you are having trouble. Please try again later. Goodbye"), the system may be unusable for this population.

3.4   Interim summary

It is fortunate for those with hearing loss that there are many effective ways to assist in their communication. These include aids that increase sound levels inside the ear and services that turn sound into the written word. Each of these methods has an effect on how an audio interface might be used, because they can change the interaction with communication devices. Designing for people who potentially use hearing assistance may seem like common sense (e.g., be clear, eliminate background noise), but knowing exactly what the users need can help avoid serious design pitfalls early in the process of display development.

3.5 Human language

Thus far, sounds have been discussed in a general sense, in terms of their perceptibility. However, language is a much more complex type of "sound" due to the human ability to consider context. We have mentioned that when older adults have hearing loss, it is often for higher frequencies of sound. In speech, these frequencies correspond to basic speech sounds (phonemes) such as *puh, kuh, sss, tuh, sh, ch,* and *th.* Even younger listeners sometimes confuse these sounds when on the phone or listening to a recording due to compression and frequency sampling.

In many instances, it is possible to perceive the "sound" of a word, even when that word was not actually spoken or heard. Consider what happens when a song has a word removed (e.g., typically due to profanity). In many cases, people state that they still "hear" the word. However, if the sound is isolated from the context around it, it becomes just a sound (such as "click"). This illusion is very strong and speaks to our ability to "fill in" missing information based on context.

The graph on the upper left of Figure 3.4 shows a waveform of a male saying "Press one for balance information." The y-axis shows amplitude, or sound pressure level. This is important because the ear hears via pressure change. Also shown in the waveform is that humans do not finish with one sound before beginning the next. As an example, when saying "I have to go to the store," most people will find themselves saying "hafta" rather than pronouncing "have" and "to" with a pause in between. For comparison, when saying "I have two keys," there is generally a distinction between "have" and "two," despite being a homonym of "have to."

This exercise partially demonstrates why people with hearing loss may have trouble with synthesized, computer-generated speech. In all but the best speech generators, these unspoken rules of pronunciation are not followed. What results is the unnatural sound of a text-to-speech engine. To understand just how much computer generation can change a voice requires a frequency analysis.

Figure 3.4 and 3.5's waveforms have smaller charts inside of them representing the frequency analysis. This analysis shows the relative amplitudes of the different frequencies that make up the complex waveform shown above each. In the analysis for the human male, most of the sound falls between 86 Hz and 10 kHz, with the loudest sounds being near the low range of 86 Hz.

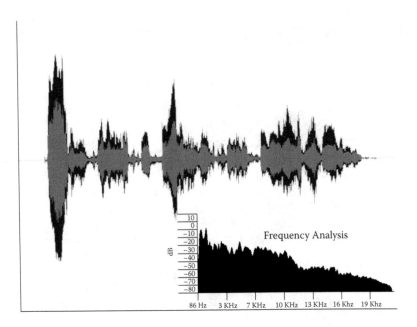

Figure 3.4 Waveforms drawing of a male speaker saying "Press one for balance information." The *y*-axis indicates the frequency of each sound wave, while the time to say the phrase is noted along the bottom of the figure. Graph shows "dB of attenuation" where "attenuation" is a negative value in decibels that describes a reduction of amplitude from the largest possible amplitude in the system.

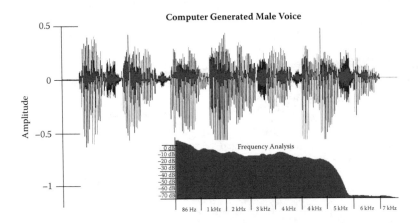

Figure 3.5 Sample of computerized speech saying "Press one for balance information" to illustrate the differences in waveforms and frequencies from the human voice.

The frequency analysis for computer-generated voices is very different from the human voices (see Figures 3.4 and 3.5). For computer-generated voices there are NO sounds outside the 7000-Hz limit. This waveform was generated using the text-to-speech engine provided by Microsoft for accessibility purposes. Unlike the human voice, the computer "voice" does not extend to the range of a real human.

3.5.1   Prosody

Prosody is a characteristic of nontonal languages that refers to the rise and fall of speech during communication.

Press one for balance information

A poor text-to-speech synthesizer would pronounce each syllable similarly in this sentence. However, a human voice would rise and fall at different points in the sentence, to stress important concepts and draw attention away from filler words.

We present two rules for using prosody in displays for older adults:

1. Use prosody. Prosody, especially when slightly exaggerated, aids in older adult comprehension of speech.
2. Do not greatly exaggerate prosody. This is commonly referred to as "elderspeak" and is disliked by older users. Overly exaggerated prosody sounds similar to "baby talk" that demeans your users. It also can trigger a stereotypical response where older users *act* older when they perceive that they are being stereotyped.

3.5.2   Speech rate

Older adults are challenged more than younger users by fast speech. Slowing speech in an audio display helps older adult comprehension; however, a number of caveats accompany this statement. There is no simple rule for what speech rate will produce the best comprehension; it depends on the speaker, the topic, the complexity of the content, the familiarity of the content, and numerous other variables. A designer might be tempted to just slow all speech as much as possible, but this too REDUCES comprehension by older adults. In other words, each display likely has an optimal speech rate that can be discovered by testing. A slightly slowed rate is probably a good start for any aging-friendly interface, but until it is tested the optimal rate will not be known. A starting point is to time the speech and make sure it does not exceed 140 words per minute, but do not depend on this rate to predict the perfect speed for a particular interface.

3.5.3   Environmental support

Environmental support consists of additional information made available during interaction with a display. In many cases, this additional information reduces the load on auditory memory by providing visual information as well. For example, a user must keep a hierarchical menu in memory when choosing options, but when that same menu is also displayed visually the user can reference the options or even plan the desired "path" before interacting with the auditory interface. Adding visual support to an auditory menu has been shown to improve older adult performance in research studies, exemplified by the card in Figure 3.6. These can be passed out with medications at pharmacies or distributed at banks or other businesses that use voice menu systems. These "cards" can be accessed virtually via websites, but this will only apply if the user population has known access to the web and the visual support card is easy to find.

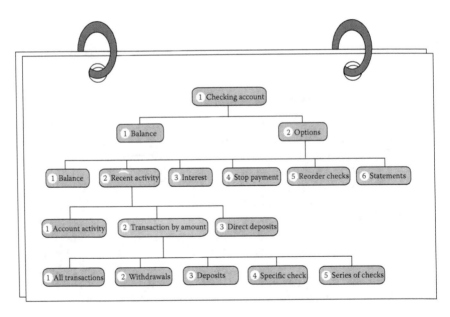

Figure 3.6 Graphical aid for a telephone menu system. (Adapted from Czaja, S. and Sharit, J., *Gerontechnology*, 2(1), 88, 2002.)

3.6   Interim summary

A wide variety of changes can occur to hearing. A good auditory design considers both the physical changes in sound perception and the cognitive changes in the comprehension that comes from initial perception. Keeping informational sounds above background noise requires a study of the display environments. The loudness of a sound is truly individual, but can be approximated through the sound pressure levels (dB) and frequency of the sound. Older adult sound and speech comprehension improves when it is in context, slowed, prosodic, and occurs at frequencies typically maintained in the aging ear. When hearing loss is severe enough that users wear an aid, consider how those aids interact with the interface.

3.7   Designing audio displays

3.7.1   Voice

Choosing the appropriate computerized voice is not a simple task. Generally, synthesized voices are not as comprehendible to older adults as human voices. Perhaps surprisingly, this may have less to do with the perceptual qualities of the synthetic voice, such as volume, and more to do with the age-related decline of short-term memory (Chapter 4). In one study, older adults had more trouble comprehending synthetic voices in a memory task. (The younger adults performed worse with the synthesized voice as well; however, the decrement was differentially larger for older adults.) This is not to say a computerized voice cannot be used; however, it should be advanced enough to mimic the prosody of human speech rather than simply turning written words into audio.

In general, designers choose female voices for voice menu systems. When designing a system to be used by older adults, it is important to make sure the voice takes advantage of the middle range of human speech and to avoid high-pitched female voices. To take advantage of the frequencies least likely to be affected by hearing loss, a lower than "average" female voice should be used. Generally, make sure all speech to be included in the voice display falls between 300 and 2500 Hz and 10 decibels higher than the anticipated background noise. Ideally, provide volume controls that are quick and easy to operate, so the user may adjust the volume to his or her hearing level.

3.7.2   Context

Context consists of the additional words or sounds in a display that allow for inference of any missed audio. For example, when hearing the sentence "The governor was standing behind ___ podium," many people do not even realize the word preceding podium was missing from the sentence. Context and expectation allowed them to fill in the missing audio. Context is crucial in any auditory communication that is degraded, such as a poorly connected phone line or interrupted streaming audio. Humans create context when possible and respond slowly, poorly, or not at all to items that violate expectation. There is some evidence that older adults compensate for some hearing loss by relying on context even more than younger adults. Providing context in a menu system is important for older adult users, because it allows them to fill in missing auditory information, whether it is missing due to the addition of background noise or the lack of hearing at a certain frequency. However, when context is absent or unfamiliar, they are affected more seriously than a younger ear in the same situation. For example, when calling a pharmacy, a menu that includes options for doctor's office calls ("Press 1 if you are calling in a prescription from a health care provider") may interfere with the understanding of the other options as the older adult tries to understand this unfamiliar and unnecessary option.

Context surrounds each choice as well. A sentence can be constructed in multiple ways. Consider the following sentences spoken over the telephone:

This pharmacy has a pharmacist on duty until 7 p.m. and is located at 4754 Capital Boulevard in Raleigh.

versus

Pharmacy open until 7 p.m., at 4754 Capital Boulevard, Raleigh.

The difference between these two sentences is the amount and order of information. The first sentence puts undue strain on auditory working memory as the user tries to create meaning while waiting for the important parts of the sentence. The second sentence prepares the user to expect, correctly, the important information in the sentence with words such as "open" followed by a time and "at" followed by an address. One research study used these different sentence types to study how well older adults used a car navigation system and found faster response times and higher understanding for the second, list-format sentence. Thus, for the pharmacy interface, the designer must examine each sentence the voice menu system produces and structure those sentences so they require the fewest memory resources.

POTENTIAL INTERACTIONS TO CONSIDER

Using a list format for information can be taken too far. Provide the context necessary for your users before using list-format information options. Also, link important information together in the list. For example, in a list of multiple pharmacies, It would be better to state the pharmacy hours, followed by the address, rather than listing the hours for each, then all locations.

3.7.3 Passive voice

An unfortunately popular technique used by many designers to make a system sound "formal" is use of the passive voice. For example, a telephone menu might state: "A representative has been contacted to assist you with your call. One moment." You can tell a phrase is passive voice when the person/object performing the action is missing (here, the action was "to contact"). Who contacted the representative? No one can know for sure who committed the action in a passive voice sentence.

In general, the passive voice makes a sentence more difficult to understand. You can change a passively written sentence to an active one by finding or inferring the actor and placing him/her/it in the subject of the sentence. For example, "Our system has contacted a representative to assist you. One moment."

3.7.4 Prompts

Prompts are an important component of auditory displays. A prompt consists of an auditory reminder for a response. In many cases, prompts consist of repetition of previously presented audio. A general failure of many auditory displays is a lack of prompts. Whether due to inattention, distraction, or forgetting, all options in a menu system are not always heard. The only way to hear these options again is to repeat the entire list. When the user is asked a specific question, such as "enter your prescription number," a long delay often results in an admonition and repetition. Consider as an example, "Your entry could not be read. Please enter your prescription number again, using the touchpad on your phone. Press the number sign when you are finished." Repeating a long entry can be just as difficult to understand the second time around as it was the first. One option is to create shorter prompts if a user doesn't appear able to make a response. Use natural language such as, "remember, press 1 for your prescriptions, 2 to talk to the pharmacist, and ...."

3.7.5 Number and order of options

The number of options should be kept to four or fewer, but at the same time the menu depth should not exceed two levels. It is a difficult task to create a menu that fits both these simplification requirements, but a thoughtful design and usability testing can often simplify a menu beyond the designer's first inclination. For example, many businesses have menu options organized in a way that fits their business model, rather than the expectations of their users. A typical bank menu is shown in Figure 3.7 to illustrate a business model of menu organization. Notice the legal statements about recording and monitoring and the long list of options corresponding to different bank departments.

Welcome to the TrustedBank center.
   (200 ms pause)
We look forward to helping you.
   (200 ms pause)
If you are calling for a loan advance this service is conveniently offered through
   our website voice response system.
   (200 ms pause)
Please choose one of the following options.
   (200 ms pause)
Your call may be monitored for quality assurance or training purposes.
   (200 ms pause)
For information or services on your existing account press 1.
   (200 ms pause)
For assistance in obtaining a salary advance through our automated services please
   press 1.
   (200 ms pause)
To apply for a loan for information on our new car buying services or if you have loan
   related questions.
   (200 ms pause)
For mortgage information or to apply for a mortgage please press 2.
   (200 ms pause)
For assistance with all other services please press 5.
   (200 ms pause)
To repeat this menu please press 9.

Figure 3.7 Example voice menu system for conducting bank business via the phone. This system was based on an existing bank menu and is organized into categories meaningful to the bank rather than the customer.

3.7.6   Alerts

Alerts should be distinct from background noise as described earlier in this chapter. Other rules of thumb for alerts are that they need to be of lower frequencies than alerts used for younger users, and an alert that uses frequencies between 500 and 1000 Hz is usually discerned by older users. Further, alerts that pulse (start then stop and start again) are more detectable than continuing tones.

3.8   In practice: the auditory interface

The textual representation of the menu shown in Figure 3.7 appears to be a very simple menu, certainly more simple than some of the nine-option menus some companies offer. However, this menu becomes deceptively complex in an audio format. Remember, the listener cannot glance back to any part of the menu that he or she misses, and must hold each option in memory while comparing every new option to find the "best" selection to complete the task.

In this menu, the user is greeted and offered a positive message. What follows should be either an instruction with how to proceed in the system or the most common choice. Here, the user is directed for a very particular activity—a loan advance (and probably not the most common option chosen)—to visit a website. The wording of this information is lengthy and confusing, and there is little information on how to access the website or what one should do with no internet access options. This first option sets up confusion and delays the understanding of subsequent options and commands. However, an audio menu cannot be paused to let the user mentally catch up.

The next information is a command to choose an option; however this is not directly followed by options. Instead, the listener is informed about their privacy rights. This is another interruption in user expectancies for the system. This is followed by a very typical menu of choices organized in a way that is useful to the bank.

However, how a bank organizes choices (by departments or their computer system) is probably not how a user organizes them. These general categories defined by the bank are: User Account, Salary Advance, Loans, Mortgages, and Other. If it is true that users think of their mortgage as being separate from a "loan," then it would make sense to list the part (mortgage) before the whole (loans) to keep users who think of their mortgage as a loan from choosing "loans" before they hear the mortgage option.

A more useful order would be to group the portions of this menu into categories: rhetorical information, instruction, and responsive information. All rhetorical information (welcome, thanks, privacy, etc.) belongs up front. Be cautious, however, because lengthy rhetorical information can produce inattention in the user, and they may tune out for the instruction and responses.

The following steps constitute one example of a redesign and testing plan.

Step 1: Make a list of all options currently offered or desired in the phone system.

Step 2: Examine previous phone system data and select the four most commonly chosen options.

Step 3: Create representative tasks for most common options and for least common options.

Step 4: Recruit older users and perform a card sort with all options. Have users write the expected functions under each option. What kind of functions and information do they expect to find under "Account Options"?

Step 5: Compare the number of groups and options within each group to the four most commonly chosen options.

Step 6: Create new interface with top four options, with user-defined functions under each option. Include other top level options under "Other."

Another design recommendation is to include natural language triggered by user responses. For example, if a user presses 3 or says "Loans," the response from the system could be "Ok, you said loans, right? Let me get that." (The system should listen for a "no" at this time.) This allows the user time to think and provides environmental support by reminding the user of the next step. This is desirable despite the time it adds.

The redesigned menu in Figure 3.8 shows significant improvements over the first system. This menu offers more options (7), but they are presented in a manageable way. First, the menu offers voice response and

---

Welcome to LocalBank!
To reach an extension, press the pound key.
    (200 ms pause)
You can always say "representative" to reach a customer representative.
    (200 ms pause)
    Please say or enter your customer number. You can also say "I don't know it" or
    "I'm not a customer yet."
    (2s pause)
    (if no response, prompt)
If you can't remember say "I don't know it" or press 1.
    (500 ms pause)
    "I'm not a customer" or press 2.
    (2s pause)
My apologies, I wasn't able to verify your account.
    I'll need to ask you a few questions so I can transfer your call.
    (100 ms pause)
Is your question concerning: insurance, investments, saving, checking, credit card, loan...
    (if user says "loans")
That was loans, right? ("Yes", "no")
    (200 ms pause)
Which account type would you like: Auto, home equity, mortgage, or personal. For all
    others say "other."
    (200 ms pause)
Or just say the name of a different account type. You can also say "more options".
    (If no response, prompt):
Please tell me the type of account you would like. You can say "new account" or press 1.
    (100 ms pause)
For quotes press 2.
    (100 ms pause)
Rates 3.
    (100 ms pause)
Financial advice 4.
    (100 ms pause)
Pin and fraud services 5.
    (100 ms pause)
Repeat 6.
    (200 ms pause)
To go back to the main menu press star 6.

Figure 3.8 Redesign of the telephone menu system to increase clarity, reduce cognitive demands, and categorize options so they fit the goals of the user.

monitors for response during presentation of the options. If the system thought the user said "loans," it replies with "That was loans, right?" If the user then says "no," the system repeats the original menu with a natural language introduction. "Ok, let me say the options again. Insurance,...." The system offers an explanation for its actions that prepares the user for a response (and prepares them for the result of their response), such as "I'll need to ask you a few questions, so I can transfer your call."

Second, notice that the menu changes based on nonresponse. Rather than repeating the same options that produced no response from the user, the interface tries different tactics. If no voice responses occur, the system offers button press options, but does not clutter the initial interface with these less natural inputs. Last, notice how the options with button presses change as they progress down the line: the first two options include extra information: "You can say 'new account' or press 1. For quotes press 2." Then the reminders to say or press disappear, as the user is only interested in the options. This is a nice implementation of menu simplification via natural language and a good example of how to move from overall context to list format.

The benefits of such a menu are many and extend beyond the hearing chapter of this book. Such improvements are helpful for working memory, language comprehension, and decision making, as discussed in Chapter 4.

3.9   General design guidelines

Based on the previous sections, we provide here a list of general design guidelines that can be used to improve the design of auditory menus.

- Calculate loudness levels.
  - Consider potential background noise.
  - For tones, use low-to-mid-range frequencies.
  - For tones, use pulses of sound rather than sustained frequencies.
- When designing a display device, consider physical proximity to the ear and interactions with hearing aids.
- Avoid computer-generated voices.
- Use prosody.
- Provide succinct prompts.
- Provide context.

3.10   Suggested readings

Berkowitz, J. P. and Casali, S. P. (1990). Influence of age on the ability to hear telephone ringers of different spectral content. *Proceedings of the Human Factors and Ergonomics Society Annual Meeting*, 132–136.

Casali, J. G. and Gerges, S. N. Y. (2006). Protection and enhancement of hearing in noise. In R. C. Williges (Ed.), *Reviews of Human Factors and Ergonomics Volume 2* (pp. 195–240). Santa Monica, CA: Human Factors and Ergonomics Society.

Cavender, A. and Ladner, R. E. (2008). Hearing impairments. In S. Harper and Y. Yesilada (Eds.), *Web Accessibility: A Foundation for Research*, (pp. 25–35). London: Springer-Verlag.

Czaja, S. and Sharit, J. (2002). The usability of telephone voice menu systems for older adults. *Gerontechnology, 2*(1), 88.

Dillion, H. (2001). *Hearing aids.* Turramurra, Australia: Boomerang Press.

Fletcher, H. and Munson, W. A. (1933). Loudness, its definition, measurement and calculation. *Journal of the Acoustical Society of America, 5,* 82–108.

Glasberg, B. R. and Moore, C. J. (2004). A revised model of loudness perception applied to cochlear hearing loss. *Hearing Research, 188,* 70–88.

Huey, R. W., Buckley, D. S., and Lerner, N. D. (1994). Audible performance of smoke alarm sounds. *Proceedings of the Human Factors and Ergonomics Society Annual Meeting, 147–151.*

Rossing, T. D. (2007). *Springer Handbook of Acoustics.* New York: Springer Science and Business Media.

Schneider, B. A. and Pichora-Fuller, M. K. (2000). Implications of perceptual deterioration for cognitive aging research. In Craik, F. I. M. and Salthouse, T. A. (Eds.), *Handbook of Aging and Cognition* (2nd ed., pp. 155–219). Mahwah, NJ: Erlbaum.

Skovenborg, E. and Nielsen, S. (October, 2004). *Evaluation of Different Loudness Models with Music and Speech Material.* Audio Engineering Society 117th Convention, San Francisco, CA.

Smith, R. A. and Prather, W. F. (1971). Phoneme discrimination in older persons under varying signal-to-noise conditions. *Journal of Speech and Hearing Research, 14*(3), 630–638.

Smither, J. A. (1992). The processing of synthetic speech by older and younger Adults. *Proceedings of the Human Factors Society Annual Meeting, 190–192.*

Timiras, P. S. (2007). *Physiological Basis of Aging and Geriatrics.* New York: Informa Healthcare.

Wingfield, A., Tun, P. A., and McCoy, S. L. (2005). Hearing loss in older adults: What it is and how it interacts with cognition. *Current Directions in Psychological Science, 14*(3), 144–148.

Zhao, H. (2001). Universal usability web design guidelines for the elderly (age 65 and older). Retrieved from http://otal.umd.edu/uupractice/elderly

chapter four

Cognition

Cognitive requirements may be the most often overlooked variable in display design. This is likely due to two reasons. First, the cognitive effects of an interface on behavior do not appear as measurable as perceptual effects; a person can easily state that they cannot read text due to visual acuity, but cannot easily explain why a set of hierarchical menus is confusing. Second, we have observed that when a display is poorly designed for cognition, users tend to blame themselves rather than the interface. Yet designing a display that fits the mind as well as the eye, ear, or hand is crucial to user success with that display.

I use a display daily that causes errors and time loss through poor cognitive design—an online teaching website that delivers materials and tests to students. This site has three modes: build, for creating and adjusting settings on the site; teach, for calculating and entering grades; and student view, for checking how a student will see the site. The display is nearly identical in each mode, but the allowable functions vary by mode. Although each word on the screen is easily read, the icons are differentiable from one another and large enough to easily click, and the content is laid out so I can see it clearly, I am constantly frustrated by attempting an action in the wrong mode and not understanding why a function I recently used appears to no longer be an option. Further, switching modes always returns the display to a home screen, so if I am many screens into a task before realizing I am in the wrong mode I must start over. Mode errors are a single item in the long list of ways cognition affects display use. I am fortunate in that I can analyze the problems with this display, but have heard my colleagues blaming themselves for errors in its use. We discuss many of these common problems and their solutions in this chapter on the cognitive abilities that change with age—both for better and for worse.

4.1   How cognition changes with age

An effective display is one that helps users in completing their goals with as little confusion and error as possible. The importance of understanding cognitive processes when designing displays is evident when the display appears relatively uncluttered, high contrast, and easy to *see* but yet is difficult to *understand*. Consider the extreme example of a road sign written in a nonnative language. In this case, the limiting factor is beyond sensations and perception (you can clearly see the road sign ahead) but in the realm of cognition: comprehending, remembering, and deciding.

Just as physical capabilities and limitations change with age, so do cognitive capabilities and limitations. When designing displays and user interfaces for older adults, it is important to understand their specific cognitive capabilities and limitations. Psychologists use the word "abilities" to refer to elemental aspects of cognition that, when combined, allow people to perceive, comprehend, and act on the information around them. Cognitive abilities include working memory (the ability to remember items for a short period of time while engaged in other tasks), spatial abilities (the ability to interpret abstract models or maps of an environment), and perceptual speed (the rate at which one can perceive and process information). Abilities are those basic elements of cognition that help people to learn about their environment and gain knowledge and skills. A useful way to think about the full range of human cognitive abilities and skills is to categorize those into *fluid* abilities and *crystallized* knowledge.

4.1.1 Fluid abilities

Fluid abilities are those abilities needed in unfamiliar, rapidly changing situations. For example, navigating an unfamiliar city or learning strategies for a new card game requires aspects of fluid cognition. The name "fluid" comes from the fact that these abilities are important in situations that change rapidly. These fluid abilities can be broken down into perceptual speed, working memory capacity, attention, reasoning ability, and spatial ability. Working memory capacity, attention, and reasoning are closely linked abilities in that individuals high on one tend to also be high on the other two, and the converse is also true.

4.1.1.1 Perceptual speed

Perceptual speed can be measured using psychometric tests, such as the Digit Symbol Substitution test. The test indicates how fast a person can perceive and compare different visual stimuli (Figure 4.1). The test taker is asked to

□	◇	‖	Γ	Δ	\
1	2	3	4	5	6

5	2	6	4	1	5	3	4	2	6	2	1	4	3	2	5	1	4	1
Δ	◇	\	Γ	□														

Figure 4.1 Simulated item from the Digit Symbol Substitution test. The goal is to fill in the appropriate symbol in the blank box (lower panel) that corresponds to the number using the key. (Wechsler Adult Intelligence Scale, Third Edition [WAIS-III]. Copyright © 1997 NCS Pearson, Inc. Reproduced with permission. All rights reserved.)

examine a group of symbol/number pairs located on top of the page. On the page are a variety of symbols with blanks beneath them. The task is to examine the symbol, figure out what number is associated with it (by consulting the reference at the top of the page), and to write down the appropriate number in the blank below each symbol. How many numbers are correctly written in a 90-second period is the digit symbol score. Although this test seems contrived, this elemental ability, combined with others, underlies many different tasks, from dialing a telephone number from a telephone directory listing to quickly deciphering the rules of a new card game.

One persistent finding in the aging literature is that fluid abilities decline with age. Perceptual speed tests clearly show this pattern. Figure 4.2 shows a plot of scores on the digit symbol substitution test as a function of age. A lower score on that test (fewer completed items) indicates a slower perceptual speed. Each dot indicates one person's score on the test. What is immediately evident is the amount of variation in digit symbol score across age decades (the vertical range of dots). The plot generally shows that people who are older tended to score lower on the digit symbol test. The negatively sloped line going through the dot cloud shows the line of best fit (the general direction of the dot cloud).

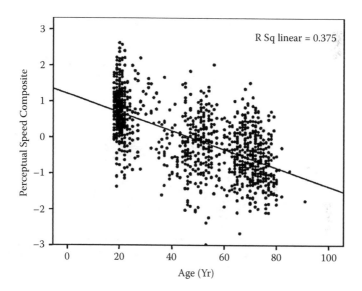

Figure 4.2 Scatter plot of perceptual speed by age (digit symbol substitution is one measure). Line indicates best-fit regression line. (Data in this figure are from the CREATE project as described in Czaja, S. J. et al., *Psychology and Aging* 21(2), 333–352. 2006. Thanks to Neil Charness who developed the figure. With permission.)

4.1.1.2 *Working memory capacity*

As the name suggests, working memory is the ability to remember information while also working on another task. Although the term is often confused with the concept of short-term memory, working memory is different because it emphasizes the ability to remember items *while also engaged* in another task. One classic example is to remember a telephone number while talking on the telephone. Trying to remember the telephone number is the main memory task, but the secondary task of conversation requires planning, memory, and comprehension.

Working memory is often measured by tests that simultaneously require retention and processing of information. In laboratory tests of working memory, aging is associated with a reduced capacity, or span, to remember items for a short period of time when also engaged in other tasks. These working memory differences by age have pervasive effects in a variety of tasks related to the use of displays. For example, the ability to comprehend and understand written or spoken textual material is affected by working memory capacity.

When comparing working memory performance between younger and older adults, the differences in capabilities are dramatic at the group level (Figure 4.3). However, even within younger adults there is substantial

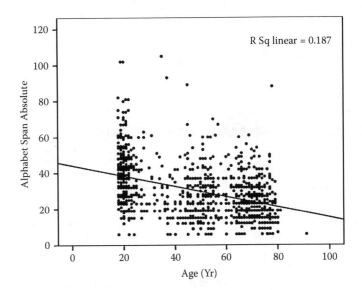

Figure 4.3 Score on alphabet span test (a measure of working memory) by age. (Data in this figure are from the CREATE project as described in Czaja, S. J. et al., *Psychology and Aging* 21(2), 333–352. 2006. Thanks to Neil Charness who developed the figure. With permission.)

individual variation, so that designing displays that reduce demands on working memory are likely to benefit users of all ages.

Fortunately, high working memory demands may be overcome by relatively modest changes to a display. The first step is to perform a task analysis that discovers the task elements with high working memory demands. Generically, any situation that calls upon the user to temporarily hold some items in memory while also carrying out another task demands working memory. This situation is more common than it may appear. For example, in a telephone voice menu system, callers are presented with a series of options ("press 1 for customer service, press 2 for billing,…"). Callers must remember their task goal, as well as the number/option pairing, while simultaneously listening and comprehending the spoken menu, and plan to coordinate the spoken options with their current goals.

A well-known design guideline is to keep the number of spoken options below seven to accommodate short term memory limitations. This guideline comes from the idea that short term memory could only hold between five and nine items. However, short term memory is static memory, or how much one can hold for short periods of time when not engaged in another task. In addition, the span of short term memory of five to nine items is typically for younger adults and for a single task. The range is lower for older adults, and the range for *working memory* is certainly smaller for both younger and older adults, closer to four items. The critical point is that for both young and old adults, only a very small amount of information can be held in *working* memory.

One way to reduce working memory demands of the task would be to present as much information as possible in a display so that memory is unnecessary. But this would obviously increase the visual complexity of the display, and users will waste time visually scanning the display. In the design of visual displays for consumer products, the user interface designer and engineer are constrained by the physical size of the display, which may necessitate limiting the amount of available information. However, during the task, only some pieces of information are required at certain points in time. For example, when trying to reach a destination using an in-car navigation aid, specific, directive information is probably more appropriate and useful than the location of opera houses or wine stores. The challenge in display design is to present the appropriate amount of information on an as-needed basis. Optimally, the display should present the amount of information appropriate to the current subtask the user is expected to accomplish.

4.1.1.2.1 Environmental support. The aforementioned specific ways of designing around working memory limitations are examples of the systematic process of understanding what parts of the task need to be supported by the display. These environmental supports, or information

Table 4.1 Task Analysis of a Web Browser Activity with Hypothesized Perception, Cogitation, and Motor Demands

Task	Task or Subtask	Perception	Cognition	Motor
1.0	Send e-mail to someone from a web link			
1.1	Scan the page for an e-mail link	Visual acuity	Attention, visual search	
1.2	Move the mouse pointer over the mail link	Visual acuity	Attention, spatial translation	Fine motor control
1.3	Click the e-mail link		Attention, visual search, declarative knowledge	Motor control
1.4	Enter the subject in the subject field		Attention	Fine motor control
1.5	Click the "send" button after composing the message	Visual acuity	Attention, visual search, declarative knowledge	Motor control

available in the task environment, operate by encouraging the more efficient use of resources people have (e.g., their knowledge) or by reducing the mental demands of the task. Creating an effective environmental support requires two commonly used methods in human factors and usability. The first is to create a user profile that accurately characterizes the intended user population's capabilities and limitations. Second is a task analysis with enough detail of the task to be able to determine whether the intended user's capabilities will be exceeded. A simplified example of a task analysis is shown in Table 4.1.

The task analysis shows a hierarchical list of the tasks and subtasks involved in sending an e-mail from a web page link. In addition, the demands placed on perception, cognition, and motor are described for each subtask. A glance at this table illustrates which tasks place heavy demands on particular abilities and may warrant possible redesign. In this example, the possibilities are to reduce the demands placed on attention by the extensive need for visual search or the spatial ability demands. This can be accomplished by making important task elements or icons as conspicuous as possible from the background while also reducing visual clutter. Spatial demands can be minimized by reducing the depth of the hierarchy or providing a site map that clearly illustrates the relationships between pages and folders.

Another way to provide support is to encourage the use of existing cognitive resources. Older adults have a wealth of knowledge and experience with certain topics. Taking advantage of this knowledge can compensate for limitations in memory, attention, or spatial abilities in performing computer-based navigation tasks. An example of such an approach is a tag-based interface, described in detail at the end of this chapter.

Even simple forms of environmental support can help older adults' performance. For example, telephone voice menu systems (sometimes called integrated voice response systems) allow users to carry out sophisticated computer interaction tasks using only a handset and keypad. However, the navigation of these menus is a task that places heavy burdens on working memory. For particularly deep hierarchies, the user must keep several pieces of information active in memory: their reason for calling, the hierarchical organization of the menu system, as well as the number/option mappings. Simply providing a graphical illustration of the menu options, their organization, and mappings has been shown to improve older adults' performance with these systems because it allows users to anticipate and plan their activities instead of blindly navigating options one after another. An example of such an aid was discussed in Chapter 3 (Figure 4.4).

4.1.1.3 Attention

A precise definition of attention is difficult; instead attention is easier discussed in terms of its everyday effects. When something is *attended to*, it appears clearer, is better processed, and possibly better remembered.

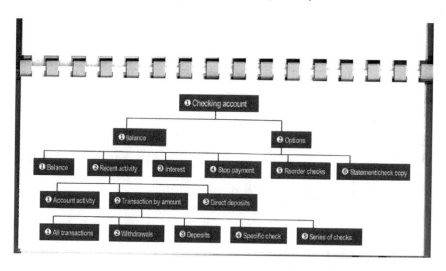

Figure 4.4 A paper-based navigational aid for the integrated voice response system (IVRS).

Depending on the user's goals in the task, attention can either be *selective* or *divided*. Selective attention is the situation where one is trying to pay attention to one thing (e.g., understanding the options on an ATM screen) while trying to actively ignore something else (e.g., traffic noise). Divided attention situations are those where one is trying to pay attention to multiple things at the same time (e.g., talking on the mobile phone while withdrawing cash from an ATM). In both situations, research studies have shown that older adults are at a disadvantage compared to other age groups.

Attention is considered a limited resource. Paying attention to something in the environment means that users will be less able to pay attention to something else. To continue the resource metaphor, research has shown that older adults have a smaller amount of this attentional resource available. The functional implication is that compared to other age groups older adults may be less able to selectively attend to elements while ignoring irrelevant or undesired elements. In addition, the task-irrelevant elements have a greater distracting effect for older adults.

When a user is faced with a display, every element on the screen commands the user's attention. As mentioned in the preface, the goal of design can be to bypass age-related limitations in attentional abilities; that is, design the display so that attention is easier to allocate. The recommendation is to reduce the amount of clutter in the display and draw attention to important or frequently used functions. For example, consider two types of interaction styles: forms-based versus wizard style.

The form-based interface presented in Figure 4.5 is very structured but also cluttered. Because of the clutter, users must selectively attend to only the relevant portions at any time, while actively disregarding irrelevant parts. An alternative is to only ask for small chunks of information at a time by using a wizard-style interface (Figure 4.5). The selective attention demands are relatively reduced compared to a form, because only a small part of the whole form requires immediate attention.

Shifting or moving attention takes time. When users are presented with a display, they will often need to move their attention throughout the display (e.g., from the menu in the upper right down to the middle, and then ending in the lower right.) Figure 4.6 shows simulated eye tracking patterns on a display. The figure shows eye movements that are assumed to reflect the movement of attention. This movement of attention takes time (known as an "information access cost").

Attentional performance is greatly affected by display characteristics. The distracting nature of task-irrelevant elements in the display can be moderated by making them appear distinctly different from task-relevant information or by removing them. However, on occasion, attention is not under conscious control. Consider the example of a loud noise in the hallway while someone types in an office. The typist's attention will be

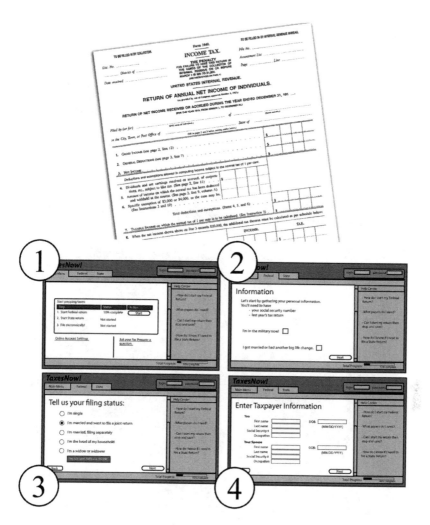

Figure 4.5 Long form compared to wizard format for data entry.

momentarily directed away from the typing task. What a user attends to in the environment, and more specifically a display, depends on the conspicuity (or how distinct it is from its surroundings.)

4.1.1.4   *Reasoning ability*

Reasoning ability is the ability to tackle and understand novel situations. It is the ability that one uses when faced with a new television remote control, visits an unfamiliar website, or tries out a new computer application without reading the manual. Psychologists measure reasoning ability using abstract tests that require test takers to determine logical sequences

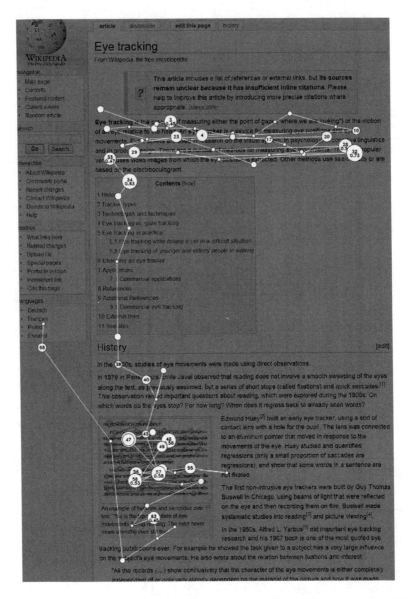

Figure 4.6 Eye-movement scan paths across a web page (assumed to reflect the movement of attention). Size of circles indicates dwell time or how long the observer looked at that point. (Image courtesy Mirametrix Research [mirametrix. com]. With permission.)

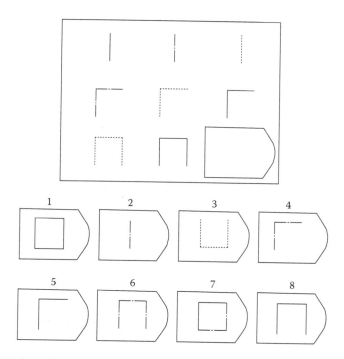

Figure 4.7 Sample test item from the Raven's Progressive Matrices Test of abstract reasoning. The test taker chooses one of the eight answer options that best satisfies the sequence in the upper panel. In this case, the correct answer is 6. (Raven, J., *Cognitive Psychology*, 41, 1–48. 2000. Reprinted with permission.)

in patterns. Figure 4.7 illustrates a sample item from such test. The task is to examine the figures on the test to discover the rule that governs the sequence of shapes and then select the correct shape in the sequence. The abstractness of the test is deliberate so that factors such as cultural background or language skill will not interfere with the results.

The link between performance on such tests and performance in a novel interface may seem distant, but they do share a common mental ability. When users pick up a new mobile phone or try to use a ticket kiosk in a foreign train station, they are carrying out mental processing similar to answering the reasoning test: examining the options on the screen and then trying out different options to discover the next logical step. Unfortunately pure reasoning ability (as best as psychologists can measure it) also shows decline with aging, with declines starting as early as age twenty (Figure 4.8).

Generally, making displays easier to use involves reducing the level of uncertainty about what to do next in the task, so that reasoning ability is less of a factor in success. This could mean being more specific about the purpose of each task step and the consequences of actions, as well as informing

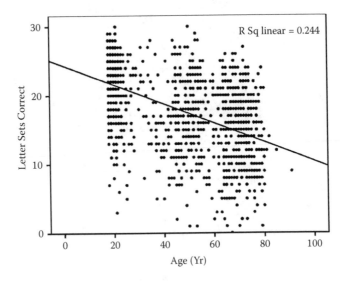

Figure 4.8 Relationship between scores on letter sets test (a measure of reasoning ability) and age. (Data in this figure are from the CREATE project as described in Czaja, S. J. et al., *Psychology and Aging*, 21(2), 333–352. 2006. Thanks to Neil Charness, who developed the figure. With permission.)

the user of their overall progress (for example, making explicit the number of steps remaining). Using icons that are less abstract and more representative of their function or task can also reduce the level of uncertainty.

However, it is rare to encounter everyday situations where one has no prior knowledge or experience and pure abstract reasoning is required. Instead, users usually always bring some amount of information or experience to these situations and use their prior knowledge to gauge expectations and guide behavior. This "mental set" is a particular way in which people approach and solve problems that is informed by prior experience or knowledge (everyday intelligence or cognition). This is why creating displays that act in ways users expect will reduce the need for reasoning ability.

4.1.1.5 Spatial ability
Spatial ability helps a person mentally manipulate location-based representations of the world. This ability is important for reading a map of an unfamiliar city or trying to orient oneself by using the navigation system in a vehicle car. In these kinds of tasks, users transform, rotate, and manipulate the physical environment in their head. People also need spatial ability when they create or manipulate mental models. A mental model is a mental representation of a physical system—a map of sorts. For example, some people have mental maps of the layout of their childhood home or neighborhood. The mental map allows them to navigate the area

quickly and may even facilitate the discovery and usage of "shortcuts" that speed navigation. In one test for spatial ability, the cube comparison test, the respondent has to decide whether the two cubes shown represent the same cube, but sitting on another face, or a completely different cube. Arriving at an answer quickly depends on the respondent's spatial abilities.

Researchers have found that spatial ability is critical in the use of some kinds of computerized interfaces and tasks such as browsing the web. For example, imagine the situation where a user browses a deep hierarchy (e.g., the Amazon.com online store). At a certain point, the user needs a mental model or map of the system so they know where they have been. The presence of the map allows users to more easily navigate the information hierarchy, because it precludes the need for the user to create their own mental versions, but such a map is harder to create for older users.

Possible Interactions to Consider

Older adults' reduced level of spatial ability may be one reason why they have difficulty with computers. The challenge is to create a display or reconfigure a task to reduce demands on spatial abilities. One straightforward recommendation is to reduce the depth of menus in favor of breadth. However, this will cause other problems because the burden that was once on spatial ability is now shifted to visual search (more breadth and less depth may mean more searching of categories). The solution is often to reevaluate the information in the display in terms of importance and frequency of use.

4.1.1.6   Interim summary of fluid abilities

Fluid abilities, subject to age-related declines, are critical in novel situations or those that change rapidly. These abilities show moderate to large age-related differences, with younger adults outperforming older adults on measures (older users are slower and have less memory capacity than younger users). Tasks that require the focusing or dividing of attention are more difficult for older users. However, age-related differences in fluid abilities such as working memory can be mitigated through the use of environmental supports.

4.1.2   Crystallized intelligence

The second category of abilities, collectively termed *crystallized intelligence* or *crystallized knowledge,* represents the sum of knowledge that one has gained through a lifetime of formal education and life experience. As the name suggests, this is knowledge and experience that may be stored in our memories about general and specific topics. This is the knowledge that one uses when faced with situations that require prior knowledge such as filling out a new tax form or insurance paperwork. In these situations, it

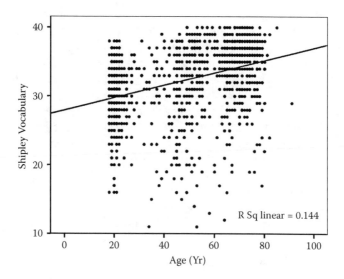

Figure 4.9 Scatter plot of vocabulary scores by age. Line indicates best fit regression line. (Data in this figure are from the CREATE project as described in Czaja, S. J. et al., *Psychology and Aging*, 21(2), 333–352. 2006. Thanks to Neil Charness, who developed the figure. With permission.)

is helpful to have a storehouse of knowledge about previous experiences that one can draw upon when needed.

4.1.2.1   Verbal ability

The measurement of crystallized intelligence is relatively straightforward. It is measured using tests of general knowledge that assess the level of factual knowledge or a more specific aspect of general knowledge: vocabulary knowledge. Tests that measure vocabulary are similar to the verbal subtest of the Scholastic Aptitude Test (SAT), which measures a person's knowledge of synonyms. Figure 4.9 shows the relationship between scores on a vocabulary test (Shipley Institute of Living Scale) and age.

The figure shows that with increasing age, word knowledge measured by the vocabulary test also increases. There is great variability within the age groups, but the general trend (indicated by the positively sloped best-fit line) is that it does not show a decline and in fact tends to increase with age. There are many possible explanations, with the simplest being that the older adults in this sample may have more education than the younger adults. However, based on number of years of education completed, the groups are virtually equal. The more likely explanation is that older adults are more familiar with the various nuances of word meaning, and this knowledge remains intact as people age. This means that despite losses in fluid abilities, older adults may be at an advantage when the task

```
DATE OF PRESCRIPTION:   07-31-97
DR:  Deems, J. M.                              RX: 081221

Melissa Hardin                                 REFILLS   1

                                               EXPIRES:  09-23-97

TAKE 1 CAPSULE ON TUESDAY AND THURSDAY, AT BREAKFAST
LANOXIN - 0.125 mg              60 CAPSULES
```

```
DATE OF PRESCRIPTION:   07-31-97
DR:  Cooper, H. W.                             RX: 081222

Melissa Hardin                                 REFILLS   1

                                               EXPIRES:  09-30-97

TAKE DAILY WITH MEALS AND AT DINNER
VASOTEC - 10 mg                 60 CAPSULES
```

```
DATE OF PRESCRIPTION:   07-31-97
DR:  Deems, J. M.                              RX: 081223

Melissa Hardin                                 REFILLS   1

                                       EXPIRES:  09-17-97

TAKE DAILY, EVERY MORNING AND BEFORE BED
PRINIVIL - 5 mg                 60 CAPSULES
```

Figure 4.10 Sample test from the Everyday Cognition Battery (Allaire, J. C., and Marsiske, M., *Psychology and Aging*, 14, 627–644. 1999. Reprinted with permission.)

requires them to draw upon crystallized intelligence (i.e., general knowledge or vocabulary).

4.1.2.2   Knowledge and experience

Knowledge of word meaning is only one specific indicator of knowledge. Knowledge and experience could be considered a type of everyday intelligence that can enable successful technology performance. Everyday intelligence is dependent on the application of knowledge learned formally and informally and can be measured through everyday cognition tests. The everyday cognition test samples people's ability to solve typical, realistic problems that may be encountered in daily life. Figure 4.10 shows an item from the test. In this item, the participant is asked to read and interpret a sample medication bottle label.

The everyday cognition test is meant to measure one's ability to use existing knowledge of previous situations on new problems (like interpreting a medication label). Crafting displays and interfaces such that they encourage the use of previous knowledge and experience may help older adults efficiently use displays.

4.1.2.3   Mental models

A convenient way to discuss a user's sum total knowledge on a specific topic (e.g., the web, word processing software) is in terms of his or her mental model. Mental models are discrete knowledge structures or "bundles of related information" that are developed over time. In human–computer interaction and applied psychology, mental models of complex systems are sometimes referred to as the user's conceptual model, user's mental model, device model, or system representation. In addition, although the terms analogy and metaphor represent slightly different things, for the purposes of the current discussion, they can be used interchangeably with mental model.

The opaqueness of using some technology requires users to infer procedures or operations by exploring the interface. For example, using a new ATM or new mobile phone, some people probably click buttons and select menus just to better understand the functions available. This discovery process can be significantly enhanced if the user has a mental model, if the system provides an explicit mental model, or one is provided in training. The mental model guides what the user should expect—how he or she thinks the system should behave and what it should and can do.

The suggestion to incorporate mental models in training and design is based on research that has shown performance benefits when users are given a mental model or cued to use their experience in a new way. Interface designers are acutely aware of the abstract nature of interacting with software interfaces and attempt to bridge the abstractness by using physical metaphors. One example of the use of physical metaphors is that of a physical desktop on which things could be placed (trash can, files). Documents or files were represented as icons and these files could be placed into "folders." Computer files are illustrated as paper files on a simulated desktop. For someone who is completely unfamiliar with computers, this model or metaphor allows them to infer possible actions on that file—they are identical to the possible actions on a physical file: open, close, trash, etc.

Mental models also have deeper benefits beyond immediate performance. A well-developed mental model can allow users to understand how things are happening "behind the interface." Studies have shown that when users are faced with a fictional device for which they have no experience, those given a mental model were not only faster, they were able to infer task procedures for which they were not explicitly trained. This does not mean that training a mental model is ultimately the best solution for older adults, because they have been shown to have more difficulty than younger adults acquiring new mental models. However, a mental model can be embedded into the interface via a metaphor that builds on the knowledge that older adults are likely to already have (e.g., the desktop metaphor).

One simple example that incorporates a specific mental model in design is the Jitterbug Mobile Phone (Figure 4.11). The phone design and plans are dramatically simplified for older adults. One feature of this phone is that the display does not contain a signal strength indicator.

Figure 4.11 Interface for a mobile phone designed for older users.

Instead, the user opens the phone and checks for a dial tone. The rationale is that older users, who may have little experience with mobile phones but a great amount of experience with land-line phones, will be able to bring to bear their knowledge and be less likely to be intimidated. The mobile phone no longer becomes advanced technology but merely a cordless phone.

Another example of designers taking advantage of a common mental model is the use of pushpins in word processing software (Figure 4.12). Pushpins are a familiar sight in the office and are used to affix items to a location so that they do not move. Similarly, users may wish to affix frequently used documents to spatial locations so that they always have access to them.

One of the most popular physical metaphors is folder tabs. Tabs are some of the simplest and most commonly used interface metaphors on the web. Tabs afford the presentation of multiple layers of information in a way that hides unnecessary information but provides easy access to

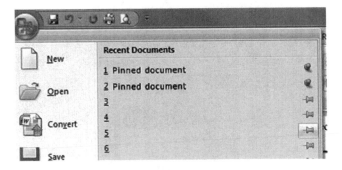

Figure 4.12 The recent documents portion of this drop-down menu allows users to pin documents so that they remain fixed.

separate information. One example of good use of tabs is illustrated by an online bookseller (Figure 4.13). The tabs in the upper left represent different categories of books or products. The tab metaphor is reinforced with subtle shadow effects to appear like three-dimensional folders sitting on a desk. Tabs have limits in their usefulness, however. Confusion may arise when there are too many tabs (Figure 4.14) or if the model or metaphor intended on the web page is not consistent with the use of tabs in the real world. For example, Wikipedia not only uses tabs to show information related to the current page (discussion tab), but also uses them inappropriately to invoke different modes (editing, and viewing page history; see Figure 4.15).

Figure 4.13 Good example of tabs on a web page.

Figure 4.14 The Amazon home page (circa 2000). Note the extensive number of tabs.

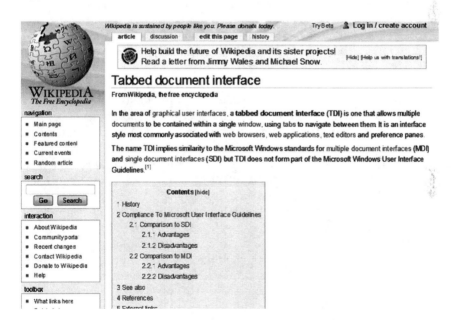

Figure 4.15 Use of tabs on Wikipedia.

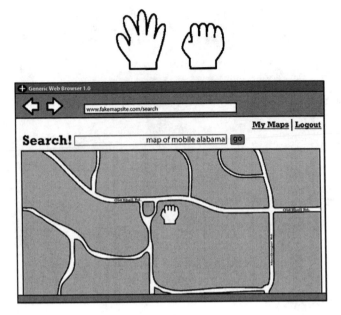

Figure 4.16 Google Maps and the two-state hand icon.

The underlying metaphor does not need to be as direct as a desktop or folder tabs, however. Subtle expectations and behaviors can be intuited by the user if the interface provides appropriate affordances. An example of such is Google Maps. Google Maps is a web application that allows users to map points of interest, search for nearby places, and navigate satellite imagery using a web browser. The website does not seem to have an overall organizational metaphor; however, the interactions required for its use are relatively intuitive and involve the hand. When the user first loads Google Maps, the mouse pointer turns into an open hand (Figure 4.16). When the user clicks the mouse button on the map in the center of the screen, the hand changes into a grasped palm to indicate to the user that they have grasped the map and can drag to move the map image. Whether this metaphor is interpreted equally well by young and older adults is unknown.

Google Maps' use of the two-state hand icon is a simple example of images that afford different user interface actions. One's understanding of a grasped palm informs the possible actions available at the interface. However, there are limitations to the efficacy of such subtle indications of possible actions. For example, the actual hand icon used in the Google Maps website is perhaps at most 10 pixels × 10 pixels in size. On a typical user display, this could be as small as a quarter of an inch. People with visual impairments may not be able to detect such subtle changes in the icon which involve around 3 to 5 pixels. In addition, even if the difference is perceived, users may not be able to detect it due to attentional issues.

4.1.2.4   Interim summary of crystallized intelligence

Unlike fluid abilities, crystallized knowledge continues to increase with age. The term crystallized knowledge, as commonly measured, represents general knowledge. Depending on the task context, users may have very little knowledge (a novice user interacting with a digital television converter box) or a great deal of knowledge (an older diabetic patient might have a well-developed model of the disease). The point, however, is that if the knowledge is present it can be used to compensate for some of the fluid ability declines.

4.2   In practice: organization of information

The organization of information is as important as the method of information presentation. The early web was dominated by simple translations of paper-based document organization onto the web with little appreciation for the differences in which people read online. When a user reads a website, the browser shows a small portion of a page (Figure 4.17). A consequence of this type of presentation is that users only see some of the information some of the time, which may lead to confusion if important elements (such as page navigation or location information) are off screen or "below the fold." This type of "periscope" presentation not only challenges working memory, it may be disorienting, particularly to new users. These usability problems can be alleviated using different strategies. For example, presenting information in smaller chunks that fit on a single page. If this is not possible, persistent navigation that follows the user is another option.

The arrangement of information can also aid the user in easily reading and finding information. For example, the layout of information on a newspaper page helps the reader's eyes scan for relevant information (narrow columns, bolded fonts) and utilizes expectations (important items in upper left, less important in lower right). Increasingly, such grid-based layouts are being used in the design of online information.

4.2.1   Page navigation versus browser navigation

Many users' first interaction with the web may have been as an informational resource (e.g., looking up information). Such applications may encourage browsing with frequent use of back and forward buttons on the browser. As users become dependent on the back button, they expect its behavior to be identical across sites. This expectancy is so ingrained, that as much as 60% of web users rely on the back button as their primary means of navigation.

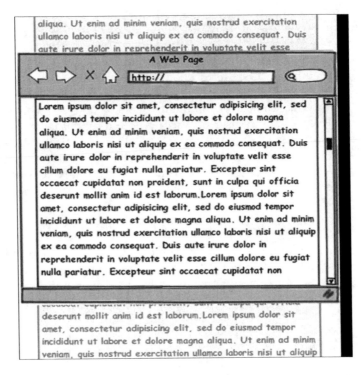

Figure 4.17 Relationship between viewing window (browser) and a web page.

With the increasing popularity of computer applications that run inside a web browser, it may become confusing for some users as their knowledge and expectations of how to use a web browser clash with their interaction expectations of a desktop application (Figure 4.18). Previously, the underlying metaphor of many web pages was that of a page that gets refreshed or reloaded when the user carries out actions. For example, when a user clicked a link or button the web page completely reloaded a new page, which sometimes involved a brief flash as the new page was loaded.

Advancements in technology have allowed web applications to mimic the behavior of desktop applications. However, confusion arises when users must shift their mental set from "web browser" to "desktop application." Actions that are appropriate for the web are no longer relevant in the web application. For example, users' general expectation of the back button is that it will take them one step back in the sequence of visited pages. However, some websites and applications disable those buttons (as TurboTax® does, illustrated in Figure 4.19), and some reprogram the back and forward buttons to take on the behaviors of a web page (as Gmail does in Figure 4.20).

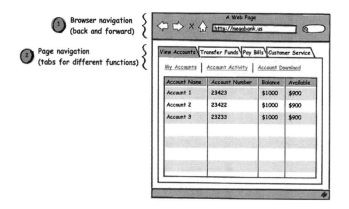

Figure 4.18 Illustration of two types of navigation. Number 1 is browser-based navigation that lets users move back and forward through their previously visited pages. Number 2 is a page/application specific navigation that allows users to move to different areas on the main web page/application.

4.2.2 Previous knowledge and browsing/searching for information

The previous sections have illustrated that aging results in a complex set of gains and losses in different abilities. An understanding of this constellation can support the design of age-sensitive interfaces that take advantage

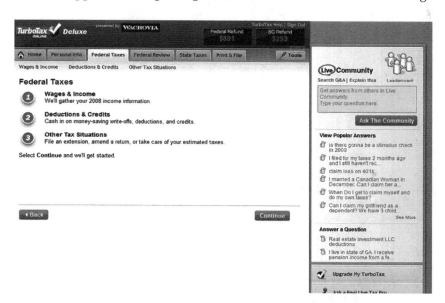

Figure 4.19 The TurboTax web application disables the browser back button. Clicking it merely refreshes the current page. The interface provides an explicit back button.

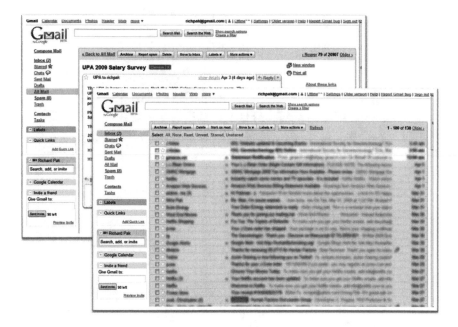

Figure 4.20 In the Gmail web application, clicking the back button (from viewing a message) takes the user back to the message list.

of gains while minimizing demands on losses. One such target is browsing and searching information-rich websites.

Websites that may be especially relevant to older adults are health information websites such as WebMD (http://www.webmd.com; Figure 4.21). These websites contain information about a wide range of health conditions, medications, and treatments. The WebMD site also presents some unique ways to browse and search the wealth of health information. The most prominent is the search bar near the top. When a user searches for a condition, he or she is presented with a typical search results list (Figure 4.22).

The unique aspect of this search results presentation is that the site uses a faceted search mechanism. In other search results lists, such as with Google Search, the user must read each search result to judge its relevance to his or her query, which can be a daunting task with thousands of results. In a faceted search interface, the search results are categorized into global categories at presentation time. For example, in Figure 4.23 a search has turned up 6000 results. Instead of examining each, the searcher can then select a facet (info type, health topic, category, or media type). Selecting a facet results in a reduction in relevant hits.

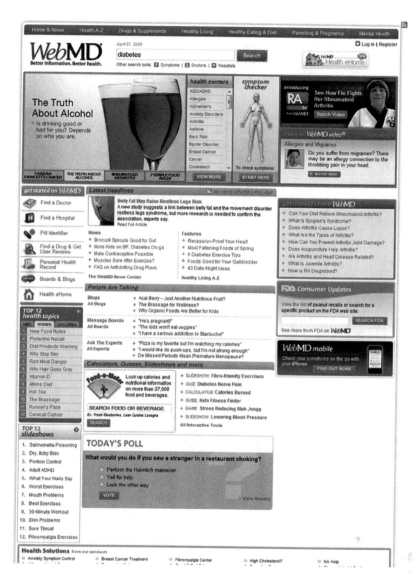

Figure 4.21 WebMD, a health information portal.

Selecting another facet results in even more specificity in the returned results (Figure 4.24).

Unlike folders, categories are not fixed and determined in a faceted display, but are dynamically generated via page metadata. A document that exists in the "diabetes" category is not solely determined by such

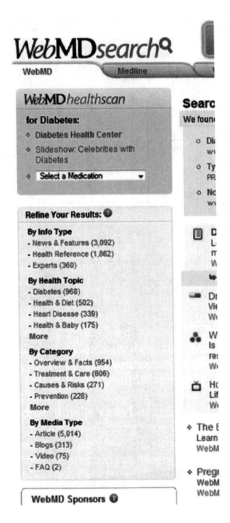

Figure 4.22 The faceted search interface.

categorization (it can also "exist" or be categorized as "treatment"). Such presentation may facilitate the user's search for specific information. For example, the searcher may not be interested in a description of the condition but on specific treatment options. In a typical search results presentation, both results would be mixed together without an easy way to differentiate the two different types of results.

Faceted search works best for search results after a user has entered a term. Another information retrieval interface that may tap into the

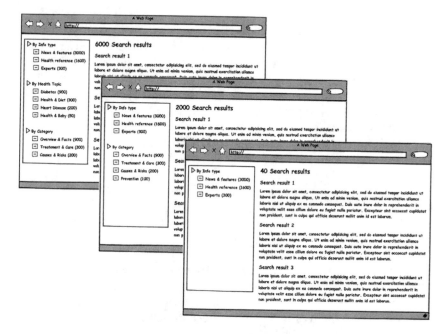

Figure 4.23 User presented with all relevant information facets.

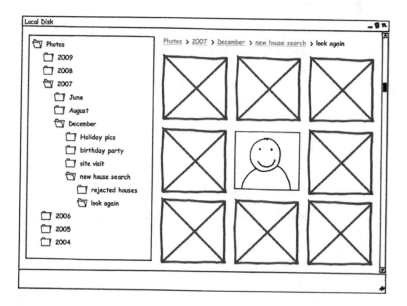

Figure 4.24 Hierarchical folder listing with folders nested within other folders.

strengths of older adult cognition is one that relies on words and concepts to browse instead of spatial metaphors or hierarchies. Interfaces that use spatial metaphors, such as folders, seemingly are a positive design decision. For example, many websites carry the file/folder metaphor to its logical end by organizing web pages (files) into hierarchical folders within folders. However, when the hierarchy gets more than a few levels deep, users' performance can be negatively affected, especially if the desired information is located in a deep level. Navigating deep hierarchies taps into abilities that are similar to those abilities useful for navigating a foreign city. When older adults are put into situations where they must navigate an information hierarchy, they invariably take longer than younger adults and make more errors.

One way to alleviate the demand on spatial ability is to shift the interface from a hierarchical file/folder metaphor (which is one source of spatial ability demands) to one with a flatter metaphor: tag-based interfaces. Tag-based interfaces are those that organize content by metadata such as dates, file types, ratings, or other user-entered keywords. The metadata (e.g., keywords) then become the actual access interface. The benefit of this type of presentation is that users no longer have to worry about the spatial "location" the information is initially located in. For example, a very organized user might organize his or her digital photographs by the year they were taken, the month, and by descriptive topic.

If the user were interested in finding a particular image of a house taken in 2007, he or she would need to drill down the hierarchy illustrated in Figure 4.24. The desired image only exists in that location (the folder titled "look again") and cannot be found elsewhere unless it was deliberately replicated in multiple places. The same picture retrieval task in a tag-based interface might look like Figure 4.25. In this interface, the user only has to recall a few cues (e.g., date, topic). Clicking on any recalled cue or keyword will display all pictures with that keyword. In this example, the user can click either "house" or "2007" to see the desired image.

4.3   General design guidelines

- Provide task-relevant information only. For example, remove extra levels of navigation if they are far away.
- Provide a clear indication of where the user is in the information space (e.g., menu structure) and task (e.g., their current step and remaining steps).
- Do not overburden attentional capabilities by presenting too much information in different forms (e.g., auditory and video information

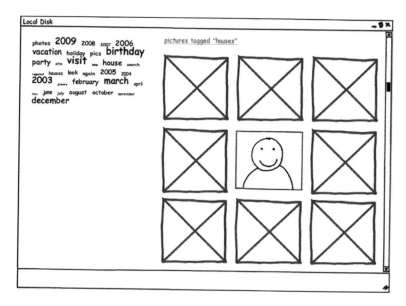

Figure 4.25 Tag-based navigation using keywords. The relative size of the words indicate number of items.

at the same time as text). This may seem counterintuitive, but remember that working memory demands have as much to do with focusing of attention as they do with the amount of material presented.

- As much as possible, use interface techniques that alleviate working memory burden (i.e., environmental support).
- As much as possible, base user interface design around a common metaphor (e.g., tabs, folders) and carry out the metaphor as long as possible.
- Provide multiple ways to access information (e.g., alternate views), because one way may not work for all users. Hierarchically based navigation may be good for some users (like tree structures in Windows File Explorer), whereas others may prefer faceted or tag-based navigation.

4.4   Suggested readings

Allaire, J. C. and Marsiske, M. (1999). Everyday cognition: Age and intellectual ability correlates. *Psychology and Aging, 14,* 627–644.

Czaja, S. J., Charness, N., Fisk, A. D., Hertzog, C., Nair, S., Rogans, W. A., and Sherit, J. (2006). Factors predicting the use of technology: Findings from the Center for Research and Education on Aging and Technology Enhancement (CREATE). *Psychology and Aging* 21(2), 333–352.

Ehrlich, K. (1996). Applied mental models in human-computer interaction. In J. Oakhill and A. Garnham (Eds.), *Mental Models in Cognitive Science*. Mahwah, NJ: Erlbaum.

Morrow, D. G. and Rogers, W. A. (2008). Environmental support: An integrative framework. *Human Factors, 50*, 589–613.

Pak, R., Czaja, S. J., Sharit, J., Rogers, W. A., and Fisk, A. D. (2008). The role of spatial abilities and age in performance in an auditory computer navigation task. *Computers in Human Behavior, 24*, 3045–3051.

Pak, R. and Price, M. M. (2008). Designing an information search interface for younger and older adults. *Human Factors, 50*, 614–628.

Raven, J. (2000). The raven's progressive matrices: Change and stability over culture and time. *Cognitive Psychology 41*, 1–48.

Stronge A. J, Rogers W. A., and Fisk A. D. (2006). Web-based information search and retrieval: Effects of strategy use and age on search success. *Human Factors, 48*, 434–446.

chapter five

Movement

An idea or desire is translated into a series of movements to become an action. For example, movement is almost always the way that we externalize thought—by whole body movements, making hand gestures or facial expressions, or vibrating our vocal chords. It is only in recent years with the ability to visualize brain activity and use brain activity to control computers that we have gained the ability to externalize thought without human movement. It is fortunate that for most activities movement is preserved well into old age, unless there is pathological disease. It may become harder to use small implements or to quickly react, but for the most part people move with few adverse effects. Technology, however, can become a stumbling block to movement performance for older adults, because many displays and interfaces require fine movements, and the consequences of an error are time-consuming and frustrating. One example is the small PDA or cell phone. A stereotype exists that older people are not as interested in these technologies as the young or that they are unwilling to learn to use them. This stereotype does not take into account the increase in initial frustration levels for someone who touches the wrong icon and has to solve what went wrong and how to fix it.

In this chapter, we cover the basics of physiological changes for movement related to age. This initial section includes some technical details about how to model movement, then addresses two age-related movement disorders: Parkinson's disease and arthritis, and how they can be designed for in a display interface. Last, we finish with a discussion of a single display, the touch screen of a phone/PDA, and how some simple design changes can assist with accurate, timely movement.

5.1 How movement changes with age

From applying the correct hand/finger pressure when picking up a wine glass (not so much that it breaks and not so little that it drops), to using a mouse to click an on-screen icon, to turning a volume knob—these are all situations where control of movement is critical. Motor control refers to the accuracy and response time of human movement. Both the accuracy and timing of movements tend to decline with increasing age irrespective of age-related disorders such as Parkinson's disease or arthritis. This can have a large impact on how people interact with displays, particularly any display with time-sensitive controls (such as ovens) or those that require accuracy (computer menus).

5.1.1   Response time

Response time, or how quickly we initiate movement, increases about 25% by the time we reach 65 years. A general rule is to provide older adults about 50% more time for a task than that task would take adults under 30. These longer response times are due to cognitive and motor changes; for example, taking longer to decide to apply the brake in a car is a function of cognitive speed, while the time it takes to move the foot and depress the brake is a motor change. These changes in movement speed are also evident in interactions with computerized displays, such as clicking on icons or browsing through menus. Interestingly, the response times for older adults are not much different from younger once the movement has been initiated. It appears that most of the additional time comes from decision time, not movement time.

Related to response time is speed of movement. Many interactions with displays require "double-clicking" for activation of some control. The control and speed of the double-click slows with age, and if a click is too "slow" many interfaces provide a different result from two clicks than they do from a double-click. One example can be found in the Windows operating system, where double-clicking a filename opens the file, while two single clicks (or a too-slow double click) makes it so the user can rename the file. This could be a very confusing and frustrating outcome for an older user who wants to open a file or start a program but is unable to double-click appropriately. The allowed time between clicks can usually be changed in the operating system settings but this is a fairly advanced option, and not many users know how to take advantage of this setting.

Fortunately, there are hardware designs that minimize the need for double-click, such as a single-click mouse. Inside the software of an interface there are numerous ways to let users choose and activate without a double-click. Of course, the double-click should probably not be disabled or cause unexpected effects as users with double-clicking would experience negative transfer (decreased performance due to being accustomed to a standard method of interaction).

5.1.2   Accuracy

Accuracy in the nontechnological world is rarely a cause for frustration (the occasional knocked-over glass or broken dish notwithstanding). Normally, if one reaches for an object and misses, the movement course can quickly be corrected and the goal still achieved. However, much of the technological world appears designed to penalize the inaccurate user.

For example, in most computer systems, inaccurately cut-and-placed text necessitates careful reselection and another placing attempt. An inadvertently dropped icon results in a return to the original location for the icon, so the user must move back to the beginning and start the task over. These small obstacles, experienced by many users of all ages, can add up to frustrate someone whose accuracy is less than it used to be.

The ability to hit a target accurately (e.g., successfully click a small icon) does decline with age, however not as much as might be assumed. Older users have been shown to be highly accurate when asked to touch targets on a screen. Older users can perform accurately, but the designer must provide adequate time to complete the movement. One heuristic is to have any target *at least* 180 × 22 pixels, but keep in mind that this is for an average resolution monitor, and pixels for smaller screens (e.g., mobile devices) will need increased size.

There are two caveats concerning retention of the ability to touch a target. First, the touch screen is a direct device. A direct device means that there is no translation required between the user's hand and the screen. Any device with gain, such as a mouse, is an indirect device, because the amount of hand movement does not exactly match the distance moved on screen. Gain is the difference in actual movement of the hand on the input device when compared to the amount of movement occurring on the display. Accuracy declines with the use of indirect devices, such as a typical mouse (though a low gain setting improves accuracy). High accuracy for older users may only be likely with direct devices. The second caveat is that this accuracy is often at the expense of response time.

5.1.2.1 Increasing accuracy

There are a number of simple ways to increase accuracy in an interface display. These techniques will increase accuracy for all users, but the techniques may slow the responses of users who do not need them. The purpose of enlarging interface elements is not only to aid those with changes in vision, but to aid accuracy in movement as well. Although as just mentioned, older users can be accurate, larger elements will help with their accuracy and speed. For example, although eliminating scrolling might be desirable, if it is inevitable, there are several elements in the scroll bar that could be made larger. One element that can be enlarged is the scroll-bar itself: Enlarging the scroll bar will help targeting. The gain of the bar can also be decreased. For example, in a long document, moving the scrollbar a small amount will move it several pages. Reducing the gain of the bar will slow down the movement through the pages, but will increase accuracy of finding a page. It has been commonly experienced by the authors that trying to find particular information using a scroll bar with high gain results in skipping over the desired information numerous times.

Other elements that can be increased in size are the arrows at the upper and lower limits of the scroll bars. Again, the scroll bar arrows are targets, so they need to be at least 180×22 pixels. If this is not possible, perhaps due to screen space on a mobile device, consider ways of organizing the display that do not involve scrolling.

Another design change that can increase accuracy is lowering the gain of indirect devices. Changing the gain allows the older user to scroll a small amount with a fairly large movement. Of course, this could be frustrating if the user needs to scroll many pages. Because of this, changing gain is a method that can benefit from adaptive interfaces. We suggest looking at algorithms to determine the intent of the user: Is the movement over a certain amount? If so, the gain can be automatically increased for that movement.

Another example of an interface change that can increase accuracy is use of "sticky" icons. Making an interface element sticky means to in some way attract and possibly hold the input device to that element. Thus, creating sticky icons is generally linked to the input device linked to that display because the input device is what acts on the element. Design of a display can be facilitated by the choice and design of input device for that display. The following are examples of stickiness for direct and indirect input devices.

When designing for a touch screen or allowing input through a pen, it is fairly common to have the cursor or input mechanism appear on the screen before the screen or pad is actually touched. The cursor will appear and move when the input device hovers over the display or touchpad. One purpose of this is often to allow a right click rather than the "left click" of a mouse that would happen if the input device contacts the screen. However, it is likely this feature will cause more problems than it solves in displays used by older adults. The phenomenon of having a cursor move from the desired active area to other areas on the screen is called "drift." Drift is often responsible for accidental activation of undesired controls, but can be controlled by requiring an explicit "tap" to any control to activate it. This is a simple fix, but the designer must be aware of the issue in order to notify programmers to disable hover activation.

When an indirect input device (such as a mouse) is used for a system, a target can be "sticky" by decreases in the gain for the device as it nears the target. Essentially, a decrease in gain allows a larger motor movement when zeroing in on a target. Decreasing the gain by small amounts (for example, from 1 to 0.9) has been shown to help younger adults quickly and accurately use a mouse to click an icon, even though users do not tend to notice the change in gain. A second option for creating "stickiness" is to enclose the target (usually an icon) in a "force field." This option allows the cursor to be drawn toward the target when it enters that field, which additionally aids achieving the target. This method is currently more programmatically complex, but is an option when accuracy on an interface is of top importance.

5.1.3 Modeling response time and accuracy

Fitts' Law is often used to calculate the expected time it takes a person to touch a target. In essence, it is a function of the size of the target and the distance to be traveled. The combination of these two things is called the "index of difficulty." It is valuable to calculate the index of difficulty for often used elements of an interface. First, we will demonstrate how to calculate the index of difficulty, then give an "age correction" factor for older users.

The Index of Difficulty is calculated as:

$$\text{Index of Difficulty} = \log_2((D/W) + 1)$$

where D = distance to target and W = width of target. The "+ 1" at the end keeps the Index of Difficulty, and consequently the predicted movement time, from ever being a negative number (which could occur with Fitts' original formulation). One advantage of using the log function is to more accurately model how users interact with the extremes of distance and target size. The index of difficulty is always less than five. Using the index of difficulty is one way to gauge *a priori* the difficulties presented by different displays; a higher index of difficulty means it will either take longer to reach a target or that users will be more inaccurate than with a lower index. Once the index of difficulty is known, it is possible to model how long it will take users to achieve that target. Fitts' Law models how long activating the target will take. One often used version of Fitts' Law (Shannon's correction) is:

$$\text{Movement Time} = a + b(\text{index of difficulty})$$

where D = distance of the movement and W = width of target; *a* and *b* are constants. The constants are calculated via regression and are composed of the intercept and slope of the data (Figure 5.1). The constant *a* refers to the intercept value, which can be thought of as the initial performance level of the user. The constant *a* can change according to experience levels. For example, a user who has used many ATMs would likely have a lower intercept value when trying a new ATM interface than a user who infrequently used ATMs. The constant *b* refers to the slope of the regression line. Once movement time is calculated, it can be converted to milliseconds by multiplying by 100. Both constants can be adjusted for an older population, with common correction being to add 75% to the younger adult time:

$$\text{Older Adult Movement Time} = \text{Movement Time} (1.75)$$

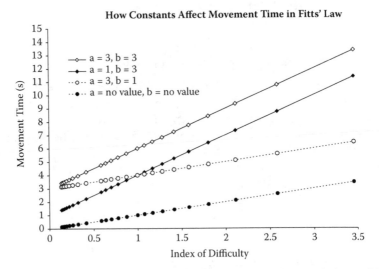

Figure 5.1 Sample movement time data as modeled by Shannon's form of Fitts' Law. Lines represent how changing the constants *a* and *b* changes the estimate of movement time. Values on *x*-axis represent different levels of difficulty.

Note that this rule of thumb correction does not change *b*, the slope, but is a correction for *a*, the intercept. There may well be slope differences for older adults, and those can be determined experimentally. A rule from Jastrzembski (2007) might be to use *b* = 100 for younger adults and *b* = 175 for older adults.

One example of using Fitts' Law to reduce movement times is through an interface element known as the pie menu (Figure 5.2). The difference

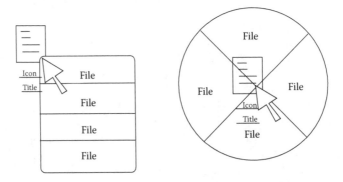

Figure 5.2 Comparison of movement times using different menus. Menu on left is a typical right-click menu for icons. Menu on right is a pie menu that reduces distance to each menu choice and increases the target size of each element.

between time to use a typical menu and a pie menu comes from the size of the target and the distance to that target. A typical menu has a short distance to some of the options, but a fairly long distance to others. The target size is the same for each. With a pie menu, the distance is the same for all options, and the target increases in size.

Another important contributor to older adult response time is the amount of time needed to make a decision about which target to acquire. This could be a choice in a menu, a particular button, or any decision point in an interface. One way to calculate the probable response time is to use a second law from the study of behavior: the Hick-Hyman Law. This law states that given a number of binary decisions (with equal probability of each choice being chosen), decision time can be calculated as:

$$\text{Decision Time} = a + b(\log(n + 1))$$

where n is the number of choices available and a and b are constants that correspond to how the choices appear in the interface and the experience level of the user. The Hick-Hyman Law requires that each choice has the same probability of selection. An example where the Hick-Hyman Law would apply is a menu with options that are equally used, varying on external conditions. For example, in a car navigation interface, a touch screen may contain choices for entering an address, selecting the destination from a list, or viewing a menu. If users are just as likely to choose any one of those, their decision time can be modeled using the Hick-Hyman formula (e.g., $a + b \log(3 + 1)$).

Knowing how these models work has contributed to many advances in interface design, such as the pie menu, and understanding the tradeoffs in target size and movement time are not considered as often as they should be. This can be seen in long forms or surveys that use tiny radio buttons that force repeated slow movement times to the target. One caveat about using these models is that they are meant to give a general idea of how quickly users can operate but are not a substitute for user testing.

5.2 Interim summary

In general, response time and accuracy are fairly predictable when people use interfaces, particularly when the experience level of the user is known. Human decision time can be calculated at the start of a design project; however, it may be of use just to consider that decisions *do* take time. Providing all possible options is not preferable to providing a limited set of options tailored to the user. Methods to improve accuracy and response time include changing target size, reducing the need for planning time, and creating interface elements that correct for inaccuracy, such as sticky icons.

5.3   Movement disorders

Although speed and accuracy of movement may change with age, these are not considered movement disorders. Movement disorders are pathological, meaning caused by disease, and are a serious threat to interface interaction. These disorders may be associated with age in that they tend to appear later in the lifespan, but they are not a result of age, and most older users do not suffer from movement disorder. This section describes movement disorders and how to design interfaces that work for multiple populations: older users with and without actual movement disorders.

5.3.1   Parkinson's disease

When Elizabeth turned 70, she had been active in a bridge club for over 10 years. Lately, however, she noticed she had more trouble arranging the cards in her hand to view, although she was still able to manage it. It took longer, and she was often irritated with herself for being the slowest to "get ready" in the group. Her playing was unaffected, however, and she and her partner won most games. However, pulling a card from her hand was increasingly a struggle, and she adapted by switching hands; her left hand was more able than her right for the first time in her life. Other activities that were affected included using e-mail and playing solitaire on her computer. She had taken a computer class that recommended playing solitaire to gain mouse and click skills, so she practiced at least one game per day. These days, however, it took her longer to play because she often "dropped" cards she was dragging before reaching the target, and the movement itself seemed to take longer. She also had a harder time typing e-mails and moving between typing and using the mouse. When she went to her doctor with these symptoms, she was told they fit the list for Parkinson's disease and was sent to a neurologist for further testing.

Parkinson's disease is a motor disorder that affects about 1% of adults over 65. Most symptoms of Parkinson's exhibit themselves as shakiness and a lack of motor accuracy and come from an imbalance of the neurotransmitter dopamine in the brain. However, Parkinson's disease actually affects the entire central nervous system, composed of the spinal cord and the brain. It is a progressive disease whose onset is often subtle, such as an arm that no longer swings normally when walking, a slight shakiness of the hands, or that an intentional movement takes much longer to complete.

In a person suffering from Parkinson's disease, part of the time increase for tasks comes from a diminished ability to perform concurrent motor movements. While a healthy 70-year-old man could place a glass in the sink while he wiped a counter, Parkinson's disease can make the hand holding the glass cease to function as the wiping motion is performed with the other hand, causing the man to drop the glass.

With time, Parkinson's progresses to much more serious lack of control, at which point an interface for Parkinson's patients would have very different requirements than a display adapted for aging users. However, if the display will be used by older adults in general, some users may be in the early to moderate stages of Parkinson's disease. Fortunately, many of the interface adaptations that make for easier use by those with early Parkinson's will also make the interface easier to use by unaffected older users.

5.3.2   Arthritis

Robert is a retired CPA who still manages his own taxes and the accounts of his wife's business. He is well versed in Microsoft Excel, as well as a number of other programs, but finds his work more and more tiring. Using the scroll wheel on the mouse to look at lengthy spreadsheets now requires him to lift his hand off the mouse and use his straight finger on the wheel; this is because he has more and more trouble bending those joints. He comes away from a computer session with pain in his hands and stiff wrists, made worse by the fact that what used to take him 30 minutes now takes at least an hour. He tries to break up this work, but the time it takes to reorient to the task when he returns is also annoying to him and adds time. He wishes he could operate a spreadsheet without small finger movements, but does not really have an idea of how this could ever be done.

Arthritis is a label that covers a multitude of disorders. Typical symptoms may include pain and swelling or stiffness of the joints. Osteoarthritis is one of the most common types of arthritis to affect older adults, because it is caused by "wear and tear" on the joints over a lifetime of use. Thus, as with hearing, arthritis is often an age-related change rather than predicted purely by age.

The main symptoms of osteoarthritis are loss of movement, joint stiffness and swelling, and a change in shape of the bones at the joint. This can greatly affect use of interfaces, from the fit of an input device into the hand, to accuracy of movement when interacting with a display.

In general, users with arthritis will have difficulty in tasks that require gripping, fluid motion of the fingers, or using specific pressure. Small knobs and close-set buttons are poor choices for users with arthritis.

5.4   Accessibility aids for movement control

Although much is known about movement disorders and their progress, there have been few attempts to study how patients with these diseases use computers. Movement disorders change more than response time and accuracy of movement; they change the nature of the interaction with an interface. For example, even when a target is acquired, people with Parkinson's tend to move from the target when clicking. The act of the clicking movement changes the position of the hand on a mouse. As mentioned earlier, movements that could previously be performed concurrently now have to be performed consecutively. Consider how many elements of typical interfaces require concurrent movement: clicking and dragging, moving a slider on a touch screen, "hovering" the mouse over an icon and right-clicking.

We provide two example designs that overcome movement-based interface problems. Both work to constrain movement and provide guides. These are not the answer to every display, but they are examples of ways designers can tackle these problems. The first of these is for mobile devices, which already present the problem of size and accuracy to those with movement difficulties. It consists of a plastic guide for touching interface elements on a PDA. These aid the user in guiding the stylus to each button (Figure 5.3). A square opening allows text entry. In this particular system, text entry is not by drawing letters, but the order in which the corners are hit by the stylus. For example, a D requires hitting the upper right corner, then lower right, then lower left. This entry system has been shown to be effective for cerebral palsy patients and younger Parkinson's patients, although the memory demands of learning a new system of writing may make the letter system unusable for older adults. However, the plastic guide is a clever way to constrain movement for those who have difficulty with control.

A second example of an assistive display is software that detects and prevents common errors. For example, given a known typing speed, an extremely fast double press of a keyboard letter can be suppressed and the mistake will not appear on the display. The same is true of other input devices, such as a computer mouse. A click that occurs while the mouse is being moved can be suppressed, because that is most likely an accidental activation of the button.

Figure 5.3 The EdgeWrite stylus guide, demonstrated on a mobile device. (Photo credit Jacob Wobbrock. With permission.)

POSSIBLE INTERACTIONS TO CONSIDER

Beware the interaction of perceptual and motor display issues. One of our suggestions is to allow increases in the size of text and targets, such as icons. However, this interacts with the suggestion to minimize scrolling (and eliminate horizontal scrolling!). The designer is faced with a difficult situation when increasing size would require scrolling, and increasing everything on the display (text and icons) may well give the user a display larger than their screen, and the offending scroll-bars will appear. If allowing size change results in hidden portions of the display, consider controlling enlargement of the display to prevent scrolling. Consider if ads or other distracters could disappear when the size increases.

5.4.1   Feedback

Feedback is considered additional information provided during or after an action to inform the user of an outcome. Feedback can be provided through all of the senses; it can be haptic, auditory, visual, or even conveyed through the human balance system (i.e., is a person upside down?). For this chapter on movement we discuss feedback provided through the sense of touch. Although many people use the terms haptic and tactile interchangeably, when we refer to tactile feedback it means through texture or touch. Haptic feedback refers to force or motion.

For design, including feedback is critical to performance with a system. What follows is a description of some options for feedback and when each is desirable. One of the benefits of feedback is that it can be included in displays for all age groups and usually not adversely affect performance for younger users.

5.4.1.1   Tactile feedback

One feature possible for input devices and displays is tactile feedback. Typically, a tactile mouse vibrates when it passes over an icon or target on the display. This works well for users with visual disabilities or blind users. Cell phones also can provide tactile feedback, electronically by vibrating when a choice is made and mechanically by having buttons that move when pressed and provide a reverberating "click" that the user can feel.

Tactile feedback can be used in various displays. A touch screen can be designed to have a screen that depresses with the feel of a button, although the hardware must be designed for this purpose rather than just a software implementation. Input devices, such as a touchpad, can have different surfaces such as light bumps or texture that indicates where permanent scrollbar areas begin. Display devices can provide force feedback, or push back against the user in certain circumstances. For example, this is used in the joystick controls for flight simulators to provide the feeling of pushing against the air. A more quotidian example is a display that provides resistance when the user attempts to perform an undesirable action (such as dragging an important system folder to the trash).

When using tactile feedback in an interface, there are some age-related changes that should be considered. Older adults have thicker skin on their fingertips and may need increased texture to achieve the same perceptual quality as younger users; this should be determined through testing. Thicker skin also makes communication through resistive touch screens more difficult because they rely on electricity flowing through the person to activate the screen. It can be frustrating for an older user when the same display that worked well a moment ago will not respond to their touch.

5.4.1.2   Auditory feedback

Auditory feedback can help interfaces that cannot offer other forms of feedback. For example, a touch screen can offer visual feedback that a choice has been made, but often cannot provide the tactile feedback of a button press. A sound can add to the visual feedback to give a better signal to the user about the action that occurred.

It is important to link any auditory feedback temporally to the motor action that caused it. For example, the buttons on a phone should sound a tone as they are pressed. Also, there should be a single feedback linked to an action. Many phones do not link the sound of the button press temporally to the sound of the signal sent across the line to dial the phone number. On such a phone, the button is pressed and a tone occurs. Briefly following, a second tone occurs, usually around the same time the second number is entered. The tones often overlap, and in this case, auditory feedback is detrimental rather than helpful to the user.

5.5   Interim summary

There are a number of movement disorders associated with increased age that should be considered during design. Parkinson's and arthritis are two pathological examples, although general slowing of response time, fatigue, and tremors are more likely to be present for older adults using an interface. Make certain to include test users who exhibit these disorders, and choose tasks that required extended interaction with the display to observed fatigue effects.

5.6   In practice: movement on a display

The example display for the Movement chapter is a typical mobile device. The device provides a touch screen with separate keypad and rocker control for input (Figure 5.4). After a brief analysis of the display we provide potential improvements for older users.

Figure 5.4 Icons representing programs on a mobile device. Icons are designed to give a common look and feel with text below each.

Before any display was created for this device the older user was already at a disadvantage, because mobile devices require a small display. Looking at a typical base menu, the "target" for direct input is smaller than 6 mm. The distance between targets is also small, crowding the display. Worse, the icons have little meaning and spill onto other (hidden) screens with little indication those other screens exist. The icons on this display have a common "look and feel" that puts aesthetics over usability. The "IM," "Contacts," and "Notes" icons all have a white foreground over a circular "shadow." All have three horizontal lines on the foreground. The only difference between these icons is the shape of the foreground container and a small person-icon on the Contacts icon. Not only are they virtually indistinguishable, taken one by one they do not convey much meaning.

Indeed, most icons are unintentionally confusing because of the "common look and feel" formula for creation. It creates an overall pleasing look, but invites disaster due to similarity of controls (like the nuclear power plants of old). Good icons should differ from each other, and this is especially important for older users. The icons (targets) here are of decent size, although the font is likely unreadable (6-point font). Vision of the text has been sacrificed for accuracy of movement (large icon targets) in this design.

One potential redesign maximizes both readability and target size (Figure 5.5). In the redesign, notice the horizontal target areas for each function were increased (at the expense of fewer targets). A scroll function exists and is allotted a large icon for that function. Scrolling would also be slowed (via gain changes), to allow more response time and to provide

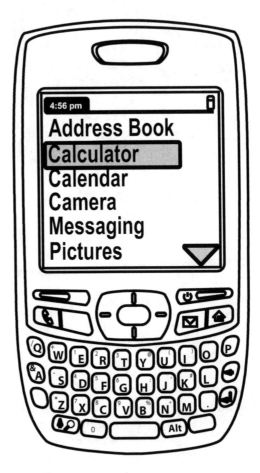

Figure 5.5 Redesign of cell phone interface for older users, taking advantage of intact verbal ability and increasing size and contrast for visual conspicuity.

more conspicuous visual feedback for older users who have lost temporal resolution.

Referring back to cognition (Chapter 4), using words instead of icons takes advantage of the general *increase* in verbal ability that comes with age. It is much faster to read a word than to recognize a picture, even when icons are good representations of their functions. For example, many phones allow the user to have a picture appear of the person calling, rather than the name in text. Cute and fun as this function may be, it takes *longer* to recognize who is calling from a picture than from a written name, even when the user chose the picture and has seen it previously. A lifetime of reading will be faster than picture mapping, particularly when the icons are so similar to one another as to be visually confusing.

5.7   General design guidelines

There is no single display or interface that solves all the difficulties older adults may experience due to movement changes or associated disorders. There are too many interfaces, interface purposes, input devices, and sizes of display for a single recipe. However, there is a higher chance of success just by knowing what issues an older population might face with the interface, planning for them, and testing the design with older users.

- Allow sufficient time for inputs.
    - Avoid timing-out operations by adding more movement interaction, such as "Click here if you need more time."
- Help guide or constrain movement for users with motor control problems.
    - Physical barriers can guide inputs.
- Offer feedback (auditory, visual, haptic), or a combination of feedback methods, matched to the environment where the display will be used.
- Gather performance and subjective experience data from users who exhibit motor control problems.
- Simplify features.
    - Reduce the number of targets by reducing the features. There is likely a feature set desired by older users. Use quantitative methods to discover those features and limit the default settings to those features.
- Rename features.
    - This advice flies in the face of most marketing decisions: have the feature named for its function, not the company who produced it. For example, if users think of "Excel" when they think

of spreadsheets, or think of the word "Spreadsheets," then use that name as the title in the display. Calling it "Microsoft Excel" is more appropriate, but changes the location alphabetically, clutters the screen, and makes a visual search for that name much more difficult (when there is also Microsoft Word, Microsoft Access, etc.). We understand this may at times not be possible due to third-party contracts and agreements. However, mentioning the function first (such as Excel versus Microsoft) may alleviate some of the problem.

- Use words.
  - A picture may be worth a thousand words, but it is a rare icon that can do this. Resist the urge to use icons just because so many interfaces have icons. On mobile devices, icons are tiny.
- Increase target size and provide accurate targeting devices, such as a stylus, rocker-bar, or knob.

5.8   Suggested readings

Ahlström, D., Hitz, M., and Leitner, G. (2006). An evaluation of sticky and force enhanced targets in multi target situations. *Proceedings of NordiCHI: Proceedings of the 4th Nordic Conference on Human-Computer Interaction: Changing Roles,* 189, 14–18.

CREATE. (2006). Fitts' Law calculator for young and old. Center for Research and Education on Aging and Technology Enhancement. http://www.psychology.gatech.edu/create/FittsLaw_d.html

Gillan, D. J., Holden, K., Adam, S., Rudisill, M., and Magee, L. (1990). How does Fitts' law fit pointing and dragging? *Proceedings of the CHI '90 Conference on Human Factors in Computing Systems,* 227–234. New York: ACM.

Grimes, D. (2004). *Parkinson's: Everything You Need to Know.* Buffalo, NY: Firefly Books.

Hart, T. A., Chaparro, B. S., and Halcomb, H. C., (2004). Designing websites for older adults: The relationship between guideline compliance and usability software. *Proceedings of the Human Factors and Ergonomics Society Annual Meeting, Human Factors and Ergonomics Society,* Santa Monica, CA, 271–274.

Jastrzembski, T. S. and Charness, N. (2007). The model human processor and the older adult: Parameter estimation and validation within a mobile phone task. *Journal of Experimental Psychology Applied* 13, 224–248.

MacKenzie, I. S. (1995). Movement time prediction in human–computer interfaces. In R. M. Baecker, W. A. S. Buxton, J. Grudin, and S. Greenberg (Eds.), *Readings in Human–Computer Interaction* (2nd ed.), (pp. 483–493). Los Altos, CA: Kaufmann.

Mandryk, R. L. and Gutwin, C. (2008). Perceptibility and utility of sticky targets. GI '08: *Proceedings of Graphics Interface,* 65–72.

McLaughlin, A. C., Simon, D., and Gillan, D. G. (in press). Title of movement control chapter. *Annual Reviews of Human Factors.*

Moffatt, K., Yuen, S., and McGrenere, J. (2008). Hover or tap? Supporting pen-based menu navigation for older adults. In *ASSETS'08: Proceedings of the 10th International ACM SIGACCESS Conference on Computers and Accessibility,* 51–58.

Paradise, J., Trewin, S., and Keates, S. (2005). Using pointing devices: Difficulties encountered and strategies employed. *Proceedings of 3rd International Conference on Universal Access and Human-Computer Interaction,* Las Vegas, NV.

Welford, A. T. (1961). Age changes in the times taken by choice, discrimination and the control of movement. *Gerontologia, 5,* 129–145.

Wobbrock, J. O., Myers, B. A., and Kembel, J. A. (2003). EdgeWrite: A stylus-based text entry method designed for high accuracy and stability of motion. *Proceedings of UIST 2003.* New York: ACM, 61–70.

Wobbrock, J. O. and Myers, B. A. (2008). Enabling Devices, empowering people: The design and evaluation of Trackball Edge Write. *Disability and Rehabilitation: Assistive Technology 3,* 35–56.

Zaphiris, P., Ghiawadwala, M., and Mughal, S. (2005). Age-centered research-based web design guidelines, *CHI 2005 Proceedings.* New York: ACM, 1897–1900.

chapter six

Older adults in the user-centered design process

A central tenet of good usability testing is to incorporate users into the design and evaluation process. It may not always be possible to include older users in the design process (although it is highly recommended), but at the very least their input should be obtained during the evaluation process. This chapter will discuss some ways that information about the capabilities, limitations, and needs of older adults can be collected and used in the user-centered design and evaluation lifecycle. We assume that the reader is familiar with user-centered design concepts and usability; hence, this chapter will focus on ways in which existing usability methods and techniques can be adapted for use with older users. More information on the general topic of usability and usability testing can be found in the recommended readings section at the end of this chapter.

6.1   How testing older users is different

Including older adults in the usability evaluation process is not that much different from including users of other age groups—one needs to have a good understanding of the target user group. Incorporating older adults in usability evaluation does involve some changes in the user-centered design process such as creating clear task instructions for the test, being aware of differing levels of technology experience, and recognizing that the testing session will take longer, and that more practice could be necessary compared to usability evaluations conducted with younger adults. The benefits of including older adults are that any recommendations generated by older adults are likely to be beneficial to users of all ages. Moreover, older adults are often enthusiastic participants and enjoy the opportunity to provide their perspectives and influence design decisions.

The organization of this chapter is modeled on the steps of the user-centered design (UCD) approach. UCD describes a general approach to the design and evaluation of interfaces which emphasizes understanding the user's needs. This is accomplished, in general, by including the user in as many steps of the design and evaluation process as possible. The process of UCD can be described in four steps described in Table 6.1. We do

Table 6.1 User-Centered Design Lifecycle

Lifecycle Step	Process/Methods
Requirements gathering (user and task)	Observation, interviews, surveys, evaluation of existing system, focus groups
Evaluation/Inspection	Heuristic evaluation, cognitive walk-through, checklists
Design/prototype/ implementation	Paper prototypes, mock-ups
Testing	Formal evaluation, performance testing

not cover every process, but highlight some that are particularly amenable to the inclusion of older users.

6.2   Requirements gathering

The requirements gathering stage of the UCD process is about under-standing the users of the interface and the problems they encounter. The preceding chapters were designed to fill in some gaps regarding older user's capabilities and limitations, but more information is needed to sufficiently describe the user (e.g., level of experience, needs that are not being fulfilled by current displays or systems, frequent or desirable tasks). The inputs for this step can be as simple as asking users about themselves and the problems they encounter (interviews, surveys), or as complex as ethnographic observation of users in their work or home environment.

Additional sources of information for gathering user characteristics may not involve the user directly. For example, it is possible to mine read-ily available data such as reports from public sources such as the Pew Internet & American Life Project or internal sources (e.g., market research). The output of this stage of UCD is a description of the user in enough detail to allow the designer or evaluator to better understand the user's capabilities and limitations in the context of the interface. Two popular ways to structure this information are user profiles and personas.

6.2.1   Age-sensitive user profiles and personas

To assist in reminding design and evaluation teams of specific user needs, it is useful to develop a user profile illustrating the capabilities, limitations, needs, and motivations of the intended users. User profiles are convenient ways of expressing the characteristics of a group of users presented in tabular form. The content of the profile is based on information from a variety of sources. An initial user profile might be obtained from general knowledge of a population, but should be refined and corrected with data from surveys, sales personnel (who have contact with customers), and any available marketing studies. User profiles are especially useful in the current context, because the design goal is to understand the unique difficulties faced by a specific segment of users: older adults.

Kuniavsky (2003) lists some example categories of information that should be contained in a user profile: demographic (age, gender, etc.), technological (level and type of experience), environment (characteristics of the typical location the system will be used), lifestyle (general attitudes and typical activities), roles (the user's primary role and responsibilities in the organization or family), goals (what the user hopes to accomplish with the product in the short and long term), needs (what the user wants from the system and why), knowledge (how much do they know about the task and product), tasks (what are the low-level typical tasks the user accomplishes with the product).

According to Hackos and Redish (1998) the user profile should be able to answer the following questions:

- What are the individual characteristics that may affect user behavior with the system?
- What do users bring with them in their heads to perform the tasks that the job requires?
- What values do they bring to the job? Are they enthusiastic learners? Are they interested in saving money? Saving time? Becoming an expert?
- What do they know about the subject matter and the tools they use today?
- What is their prior experience using similar tools and interfaces?
- What are their actual jobs and tasks? What reasons do they have for using the product?

User profiles are also useful for recruiting participants for usability testing sessions or interviews. Table 6.2 shows an example user profile table illustrating three different user groups for an automated teller machine. These user attributes are specifically chosen for the system being evaluated; others may be more relevant for other systems. Table 6.2 shows other user attributes that may be relevant for a variety of displays. The specific information about older users can come from surveys of users, interviews, or may be readily available from different sources such as existing market data.

For some design projects it may be sufficient to have user profiles that describe the capabilities and limitations of users as presented in the profile. However, some projects may require more knowledge about the user's motivations, attitudes, and importantly, how they will behave in specific situations. In these cases, a persona may be useful.

A persona is used to visualize the intended user of a system. A persona is a fictional individual based on an existing user profile. Personas are similar to user profiles in that they make explicit statements about the intended

Table 6.2 User Profile of Three Age Groups for an Automatic Teller Machine (ATM)

User Characteristic	Teens/Young Adults	Young Adults to Middle Age	Middle Age to Senior Citizens
Age	12 to 25	25 to 50	50 to 80+
Sex	Both male and female	Both male and female	Both male and female
Physical limitations	May be fully able-bodied or have some physical limitations in relation to, for example, hearing or sight Will be of varying heights	May be fully able-bodied or have some physical limitations in relation to, for example, hearing or sight Will be of varying heights	May be fully able-bodied or have some physical limitations in relation to, for example, hearing, a sight, mobility, or the use of hands Will be of varying heights
Educational background	May have minimal or no educational qualifications	May have only minimal educational qualifications	May have only minimal educational qualifications
Computer/IT use	Probably have some prior experience of computer or IT use	May have little or no prior experience of computer or IT use	May have little or no prior experience of computer or IT use
Motivation	Probably very motivated to use the ATM, especially in relation to their banking habits	Could be very motivated to use the ATM, especially if they can do their banking quickly and avoid standing in line at the bank	Could be very motivated to use the ATM, but would probably prefer to stand in a line in the bank
Attitude	Attitudes to use may vary, depending on the services the ATM offers and the reliability of the technology itself	Attitudes to use may vary, depending on the services the ATM offers and the reliability of the technology itself	Attitudes to use may vary, depending on the services the ATM offers and the reliability of the technology itself

Source: Stone, D., Jarrett, S., Woodroffe, M., and Minocha, S. (2005); *User interface design and evaluation.* San Francisco: Morgan Kaufmann. Amsterdam: Elsevier. With permission.

user; however, personas represent a smaller group of core users to such an extent that the persona can be "named." In a sense the persona represents the archetype of a user that the user profile describes. The utility of personas comes from the ease with which they are remembered by the designer and evaluator. They may also allow the evaluator and designer to predict how a user will behave in a task scenario, such as whether the user will use a feature of the display or be confused by the navigation or page layout. One way to view the persona is that it puts a name and face on what would typically be dry, aggregated, demographic data. This is often more "usable" for the designers when considering features and other design elements.

Base the persona on *data* for a typical user when creating personas of older users, because it is all too easy to stereotype aging users. Older users are sometimes described as "not leaving the house much" or "having nothing to do since they retired." Stereotypes like these rob personas of their ability to communicate actual information about a population that can be busy all day coordinating activities and social engagements. Moreover, personas are not simply verbose narratives of users; they represent a snapshot of a user's behavior patterns with enough detail to predict how that individual will react in certain situations.

One way that user profiles and personas of older adults might be different from those describing other age groups is in terms of attitudes toward technology and physiological attributes. Chapters 1 through 5 discussed the ways in which older adults are different from other age groups along cognitive, perceptual, and motor categories. This information can be used during the development of the user profiles and personas. Of course, entire age groups cannot be categorized along a few dimensions, but generalizations extracted from the literature are useful to highlight how older users are different from other users in design-relevant ways. It should be noted that some texts use the terms "user profile" and "persona" interchangeably or use one term to describe the other concept. However, both tools have identical functions: to focus design and evaluation efforts around a concrete user (instead of a diffuse "user") and to solidify assumptions about users.

6.2.1.1   *Technological demographics and attitudes toward technology*

In general, older adults have less familiarity and experience with technology compared to other age groups. The magnitude of this difference is often exaggerated by stereotypes of the typical older user. Our own studies with older users show that in terms of frequency and length of computer use or exposure to various forms of technology, age differences are not as large as one might expect. Recent census data and Pew studies also show that the usage gap is closing between older and younger users. Although older users are less frequent users of technology and may have less exposure to various forms, this is not an indication of a general resistance toward technology. On the contrary, older adults are often practical in terms of technology adoption. If the system has clearly articulated benefits, then older adults are quite willing to overcome other difficulties (e.g., cost, time to learn), although they may need support in doing so (e.g., well-designed displays and instructions). If the system does not present sufficient perceived benefit over an existing system then they may not perceive benefit in investing time to learn to use it. Lack of perceived benefit is a very different reason for not adopting technology than fear or inability to learn to use a new technology.

6.2.1.2   *Physiological attributes*
As reviewed in the Vision, Hearing, and Movement chapters (Chapters 2, 3, and 5), older adults have some limitations in physical abilities that may impact their successful use of displays and controls. Keeping these attributes in mind during design and evaluation are important and may influence the choice of input device or other display control.

6.2.2   Task analysis

Understanding user needs and motivations is important, but so is understanding the goals of the user with the system or display. Task analysis is the technique of dividing a continuous task into its step-by-step actions to identify the specific physical and cognitive demands. Each action often contains subtasks that are not evident to the expert user or designer, but are revealed in a careful task analysis. Task analysis is also useful to illustrate the complexity of apparently simple tasks. Table 4.1 from Chapter 4 is an example of a task analysis that shows the low-level subtasks required in the task of browsing the web but also shows the cognitive ability demands for each subtask. The best way to conduct an age-specific task analysis is to first understand the fundamentals of age-related change (Chapters 2 through 5), then consider those potential effects on each step in the task. For example, a task or subtask that may seem trivially dependent on working memory and therefore not noted for younger users may be near the limits of older adults' working memory capabilities (e.g., having to remember information from one screen to the next). It is critical to have a good understanding of older adults' capabilities and limitations before performing the task analysis.

6.2.3   Surveys

Surveys are useful at many points in the usability lifecycle. They may be used very early if characteristics of the user population are unknown (e.g., age ranges, level of experience) or as evaluative devices once a display has been created. Surveys are also relatively low-cost ways of obtaining information about the type and frequency of problems that users are having with the user interface. Last, they are useful for assessing user attitudes toward new technologies. Specifics of survey design are better covered in other texts (see recommended readings section) but a few general rules about creating surveys for older users include: use a 14-point font; if a scale spans more than one page, provide labels for the scale on each page; and last, provide the answering method close to the question. For example, we have seen some surveys that require the survey taker to fill in answer bubbles at the end of the survey. This is unacceptable because it requires the survey taker to remember the answer and match each question to an answer-space distant from the question, as well as the extra step of translating their answer into a letter.

6.2.4 Focus groups

Focus groups involve conducting structured interviews with users brought together to discuss several specific topics in depth. The value of focus group research comes from the depth of this discussion, which is why a skilled moderator is an invaluable asset to a focus group study. In general, the rules for groups are to have a structured list of topics to cover (about five to seven, but seven can be difficult to cover in the typical 2-hour span of a group). The people in each group should not know each other well. It is the job of the moderator to draw information from each group member; focus groups can be considered to be a number of interviews performed at once. Much more information can come out when people hear other responses to questions and discuss among themselves. The value of the focus group comes from the interaction during these "interviews." Another rule for focus groups is that each group should be homogeneous within a group with respect to the research questions. For example, on some topics it may be important to have all female or all male groups. For other topics it may be important to have all older adults in one group and younger adults in another. The heterogeneity/diversity required for the focus group analysis is obtained across groups by having multiple groups. A good resource for starting a focus group can be found in the book *Designing for Older Adults*, referenced at the end of this chapter.

Focus groups with older adults require a structure for the discussion, which is often overlooked when testers have a large number of questions about the interface or display being discussed. When conducting multiple groups there must be a way to compare experiences across sessions, and a good structure allows for this. Data analysis of research focus groups is systematic in terms of coding responses; thus a similar structure for all discussions will provide examples of focus member responses to a particular issue.

The structure and analysis of a focus group can be illustrated by a focus group study with older farmers concerning the displays on their equipment. The goal of the groups was to uncover knowledge from experienced farmers regarding dangers and safety issues on their farms. The structure developed for the groups was to specifically ask about equipment displays, tasks, time of year for those tasks, and the farmers' estimates of the most dangerous events and what made them dangerous. The moderator was looking for mentions of events that violated expectations (such as a display on a piece of machinery that provided misleading information) and various other issues, such as lack of training in reading certain displays or ignoring important display elements. As expected from a focus group, these data were gathered along with unexpected insights. For example, older farmers were aware of their declines in movement speed and adjusted their tasks accordingly, but also mentioned that completing tasks was paramount and would work into the night to finish them. In this case, the normal displays on their equipment were not visible and were ignored. These insights would not have been likely to show up in written surveys or even through observation during daily tasks.

6.2.5   Interviews

Many of the benefits of focus groups also apply to interviews. Interviews afford the opportunity to gather richly detailed information about how users go about their tasks, the conditions surrounding the task, and problems they had during the process. The benefit of an interview over a focus group is that more detail about individual experiences can be collected. Different questions can be marked as follow-ups depending on answers, and an interview can also be paired with observation of the task. The main limiting factor in the quality of data is the interviewer and the script used during interviewing: a poorly worded or vague script can result in useless data. It is important to pretest the interview questions to make sure that the participants understand the questions and that the participant interprets the question in the way intended by the researcher.

6.2.6   Observation studies

User observations can occur with a product or system of interest or consist of observing how users act in an environment where a product will eventually be used. In the second case, observational studies can contribute greatly to initial design specifications and often reduce the number of iterations required in usability testing.

6.3   Evaluation/inspection

One of the outputs of the requirements gathering stage of the UCD is an inventory of the problems that users commonly encounter in the course of their tasks with the interface. However, users may not be clearly able to articulate a specific problem or, because of a tendency to blame themselves, they simply may not perceive design-induced usability problems even as they experience them. To accompany the user-generated list of usability issues, it is helpful for the evaluator to inspect the interface to compare against a list of well-known design guidelines or heuristics.

6.3.1   Heuristic evaluations

Heuristic evaluation is a method of quickly assessing a user interface or system for its adherence to a set of predefined usability principles (Table 6.3). Heuristic evaluations are popular because of their ease of use and low cost (no special equipment is required). A set of evaluators (three

Table 6.3 Nielsen's Ten Usability Heuristics, Adapted for an Older User Population

Heuristic	Description
Visibility of system status	The system should always keep users informed about what is going on, through appropriate feedback within reasonable time. *The feedback should not only be matched closely in time to the action, but should be physically close to the action because of a potentially smaller field of view for older users and susceptibility to change blindness, especially when that change occurs in the periphery.*
Match between system and the real world	The system should speak the users' language, with words, phrases and concepts familiar to the user, rather than system-oriented terms. Follow real-world conventions, making information appear in a natural and logical order. *Natural language is especially important for older users. Keep in mind that every screen or item of feedback is helping to create a mental model of the system for the user.*
User control and freedom	Users often choose system functions by mistake and will need a clearly marked "emergency exit" to leave the unwanted state without having to go through an extended dialogue. Support undo and redo. *If a technology is intimidating to an older user population, this heuristic is even more important than for younger users. Not only do users need an emergency exit, they need to have it obvious that this exit will exist before they take a chance on an action they are not sure will be the correct choice for their goals.*
Consistency and standards	Users should not have to wonder whether different words, situations, or actions mean the same thing. Follow platform conventions. *While platform conventions are important, older users may not be familiar with them. Consistency within an interface is important, but can be confused with a common "look and feel." Display elements should be perceptually different enough that they are not confused by the user.*

(continued)

Table 6.3 Nielsen's Ten Usability Heuristics, Adapted for an Older User Population (Continued)

Heuristic	Description
Error prevention	Even better than good error messages is a careful design which prevents a problem from occurring in the first place. Either eliminate error-prone conditions or check for them and present users with a confirmation option before they commit to the action.
Recognition rather than recall	Minimize the user's memory load by making objects, actions, and options visible. The user should not have to remember information from one part of the dialogue to another. Instructions for use of the system should be visible or easily retrievable whenever appropriate. *This heuristic is critical for older users. Providing environmental support in an interface is difficult to achieve without increasing clutter and the need for visual search, but it is a goal worth iterating toward.*
Flexibility and efficiency of use	Accelerators—unseen by the novice user—may often speed up the interaction for the expert user such that the system can cater to both inexperienced and experienced users. Allow users to tailor frequent actions. *Be careful with automatic adaptation to use because changing a display from one time period to another can be confusing to the user. The convention of "don't show this again" often works well to help the user tailor actions, but things such as shortening menus or rearranging icons should probably be avoided.*
Aesthetic and minimalist design	Dialogues should not contain information which is irrelevant or rarely needed. Every extra unit of information in a dialogue competes with the relevant units of information and diminishes their relative visibility. *However, continue to use natural language and avoid shorthand via jargon.*
Help users recognize, diagnose, and recover from errors	Error messages should be expressed in plain language (no codes), precisely indicate the problem, and constructively suggest a solution.
Help and documentation	Even though it is better if the system can be used without documentation, it may be necessary to provide help and documentation. Any such information should be easy to search, focused on the user's task, list concrete tasks to be carried out, and not be too large. *Older adults are more likely than younger adults to rely on help and documentation.*

Note: Age-relevant additions are noted in italics and were created for this book.

Source: http://www.useit.com/papers/heuristic/heuristic_list.html. With permission.

to five) compare an interface against a list of usability heuristics and note any violations of the usability principles. The evaluators then gather to discuss and rank each problem in terms of severity. Heuristic evaluations can be specific to the type of system being evaluated (e.g., heuristics for mobile phone interface design, game design).

No special considerations need to be taken when using heuristic evaluations for older adults. However, extra attention should be devoted to the fact that criteria for what may constitute a heuristic violation may be different for older adults than for younger adults. For example, consider the first of Nielsen's heuristics (visibility of system status). Feedback should be given in a reasonable amount of time, but what is reasonable for younger adults may be much too short for older adults, not only because of perceptual deficits but because of attentional constraints—they may not be attending to the proper location at exactly the right time. A rich set of heuristics specific to older web users have been developed by Dana Chisnell and her colleagues. The reference to their heuristics is available at the end of this chapter.

6.4 Designing/prototyping/ implementing alternate designs

Measurement is at the core of quantitative methods of research. For a method to be quantitative, it must seek and measure evidence that can be mathematically evaluated. In other words, the collected data are ordinal (they have some ranked order [e.g., "good, better, best"]), interval (they have an order and the distance between numbers is meaningful [e.g., 10 degrees Fahrenheit versus 20 degrees Fahrenheit]), or ratio data (the numbers are on a continuous scale and have a meaningful zero level [1 lb, 2 lbs, 10 lbs]). These results can be presented graphically and tested with inferential statistics, such as measuring significant differences between groups. Because the data to come from quantitative methods are so important when considering specifics of a design, such as time or attention needed for use, tests must include a number of older users. For example, if one were testing how much time is needed for a driver to make a decision to turn (when using a navigation system), testing only younger users would provide inaccurate numbers when considering an older users' decision time and movement time.

6.4.1 Paper mock-ups/prototyping

Getting feedback from users about the layout and functionality of a display does not require specialized programming knowledge or even artistic skills. Paper prototyping is a way to obtain quick feedback about how

Figure 6.1 A paper prototyping session. A user is interacting with a rough, paper-based prototype of the interface. The user is using his finger as the mouse pointer, and the evaluator changes the "display" by moving sheets of paper. (Available on Flickr.com under a Creative Commons Attribution 3.0 Unported license. http://www.flickr.com/photos/21218849@N03/3902255728/. With permission.)

a user engages an interface in the course of a task. Paper prototyping is the process of using common office supplies to mock up rough prototypes of the display, putting them in front of users, and obtaining feedback. Because of the simplicity of this method, it is also one of the easiest ways to get information about how the interface is perceived and acted upon by users. The ease with which paper prototypes can be created can encourage frequent and iterative testing, resulting in a more refined display than other high-fidelity prototypes. Figure 6.1 shows an example paper prototype of a web page. Notice the rough nature of the prototype—it is not intended in any way to reflect the final design aesthetics, only functionality and arrangement. Relevant to testing with an older population, paper prototypes also may be less intimidating than sitting in front of a computer or other system and may encourage more open and honest feedback of the display arrangement or task structure. The participant in Figure 6.1 is walking through a task with the paper prototype, using a finger as a mouse. When the participant makes an action that involves an interface change the moderator changes the screens (either by switching prototypes or by laying down modal dialog boxes made of bits of paper).

A slight variation on paper prototyping concept is a usability technique developed by Thomas Tullis called freehand interactive design offline (FIDO). Whereas paper prototyping is meant to evaluate the layout and functionality of an interface, FIDO is a technique used earlier in the

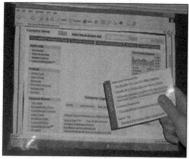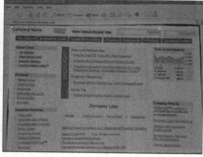

Figure 6.2 A simulated web browser window. In the FIDO technique, the user places magnetic pieces of the user interface as desired. (© UPA, 2004, Reprinted from the UPA 2004 Annual Conference Peer Reviewed Paper "Freehand Interactive Design Offline [F.I.D.O.]: A New Methodology for Participatory Design," Tedesco, Chadwick-Dias, and Tullis. With permission.)

design process to determine interface layout (Figure 6.2). The technique involves placing common interface elements on magnetic stickers and having users essentially build an interface to suit their needs.

The benefit of low-fidelity techniques such as paper prototyping (for evaluation) and FIDO (for design) are that they may present a less intimidating task for novice users. Instead of a focus on explicit performance measures (task time) paper prototyping and FIDO sessions are more like elicitations of feedback with no right or wrong answers.

6.4.1.1 Representative tasks

It is important to understand what tasks are most common, desirable, and crucial to a user being able to operate or understand a display. Developing representative tasks requires input from multiple sources, such as users, programmers, and designers. These tasks will be the tasks users discuss in formative evaluations and the tasks they perform in summative evaluations of displays. A fundamental understanding of what elements of the display are most likely to present issues for older users can provide guidance for the selection of representative tasks, and it is important that these tasks are included for testing.

6.4.2 Simulating effects of aging

Another approach to user testing for products that will be used by older adults is to attempt to simulate the effects of aging. For example, some products attempt to mimic age-related perceptual, cognitive, and motor changes in performance. This can be done in many ways, from restriction of motion through mechanical means to yellow glasses that mimic the yellowing of the lens. This approach is popular with designers who do not have access to older test subjects or cannot involve older subjects in their testing. Perhaps the most benefit to come from this approach is to allow the "experience" of the older user to occur for a younger designer. It is said that showing is better than telling, and indeed, designers who try to use a product while wearing devices that mimic age-related change are often less prone to blame the user for difficulties.

There are some problems with current devices that mimic aging that should be solved in the future. First, the actual products (usually called "aging suits") are proprietary to the companies that developed them. There are very few of these suits in existence, and one must pay the company to use the suit. Data analysis coming from use of the suit is also provided by the company, so there is little transparency or control by the researchers. Last, although mimicking cognitive change is a general goal, these devices have concentrated on the perceptual and motor difficulties that can be experienced by older adults and typically do not address cognitive changes. For example, there is no way for the current suits to mimic the inability to inhibit distracting visual stimuli.

6.5   Recruiting

The goal of recruiting older participants for a usability study is to achieve representation of the potential differences older users might face with a product or system. There are a variety of ways to recruit older participants for usability testing. Newspaper advertisements may provide access to a wide range of individuals. Independent living facilities are also possible sources. Conducting tests at the facility will provide access to a large number of participants and will likely help everyone to arrive on time as well. Sometimes a facility will charge a fee for use of a room for testing.

Involving the community in the work is another good way to recruit participants. Giving presentations on products and what one plans to test is a good time to collect names and phone numbers. These presentations can be given at community centers, lodges, or senior centers. Stress the importance of involving people of all ages in the testing process to make sure a display is appropriate for everyone. Even if the people at the presentation are not interested in participating, they likely know people who are and can pass the message along. Always bring brochures with clear contact information and an explanation of what being in a usability study will be like. Once they agree to help test a system, the most important rule we have found is to make sure that people leave feeling good about their experience. Giving them a frustrating display to disentangle can leave participants feeling as though *they* are the failure, not the display. We repeatedly remind participants that we want their opinions on the system because we are testing the system, not them. This may seem obvious to the tester, but it is not obvious to the participant, and such reassurances make it more likely that individual can be recruited for future tests.

Although these recruitment techniques may yield participants of the appropriate age range, these users may be very different from the actual users of the display. Sampling concerns affect all of the methods described in this chapter: holding a test at a corporate or university site usually restricts participation to people who still drive; recruiting from independent living results in a high socioeconomic status sample. Being aware that a sample could be biased will help in preparing materials to discover if that bias could affect experience with the display. For example, if a driving-related display were to be analyzed in a usability test then holding the test at a site participants need to drive to is likely not a problem.

6.6   Summary

In summary, all of the usability methods appropriate for younger users can be effectively used to test older user's experiences with a display. However, just as the unique capabilities and limitations of older adults need to be considered in the general design process, they also need to be considered in the user testing process by choosing representative tasks and involving older users in the formative evaluations of the display. Some of these methods are to inform the iterative design process, whereas others are meant to enable the designers themselves to be sensitive to aging issues through use of personas and aging suits.

6.7   Suggested readings

Chisnell, D. E., Redish, J. C., and Lee, A. (2006). New heuristics for understanding older adults as web users. *Technical Communication, 53*(1), 39–59.

Dumas, J. S. and Salzman, M. C. (2006). Usability assessment methods. In R. C. Williges (Ed.), *Reviews of Human Factors and Ergonomics,* Volume 2 (pp. 109–140). Santa Monica, CA: Human Factors and Ergonomics Society.

Fisk, A. D., Rogers, W. A., Charness, N., Czaja, S. J., and Sharit, J. (2009). *Designing for Older Adults: Principles and Creative Human Factors Approaches* (2nd ed.). Boca Raton, FL: CRC Press.

Hackos, J. T. and Redish, J. C. (1998). *User and Task Analysis for Interface Design.* New York: John Wiley & Sons.

Jastrzembski, T. S. and Charness, N. (2008). The model human processor and the older adult: Parameter estimation and validation within a mobile phone task. *Journal of Experimental Psychology: Applied, 13,* 224–248.

Kuniavsky, M. (2003). *Observing the User Experience.* San Francisco: Morgan Kaufmann.

Nielsen, J. (1996). *Usability Engineering.* Amsterdam: Elsevier.

Rubin, J. (2008). *Handbook of Usability Testing.* Hoboken, NJ: Wiley.

Stone, D., Jarrett, S., Woodroffe, M., Minocha, S. (2005). *User interface design and evaluation.* San Francisco: Morgan Kaufmann.

chapter seven

Integrative example
Mobile phone

Mobile phones (or "handsets") are increasingly used for a wide range of communication activities beyond voice communication. One of these functions is mobile electronic mail (e-mail). It represents an interesting test case in that the constraints imposed by design for a small screen can illustrate problems that may affect older users. The specific type of phone is known as a "feature-phone" or one that has more capabilities than just voice communication.

It should be noted that mobile phone interfaces (as well as many other interfaces) are the work of a team of people who may not directly work together (independent software developers, original-equipment manufacturers), and that many of these usability issues may be the direct result of this lack of integration. For example, the specific e-mail application discussed in this chapter may have been developed by an independent software company ("third party") to be installed on many different handset models from many different carriers, and each model can vary in button layout and placement; thus the recommendations may range from specific to general, so as to make them widely applicable.

In this chapter we examine how the mobile e-mail task might be redesigned with an older user in mind. To make the evaluation and redesign easier, we have employed a simplified user-centered procedure. We first created a user profile of a representative older adult user and focused the evaluation and redesign around his/her needs. The user profile is populated with relevant user characteristics such as name, age, experience and education, goals and motivations, and tasks.

The feature–phone interface is driven by a soft-key and directional arrow input method. Primary navigation is achieved through a four-way directional pad (d-pad) with a selection button (Figure 7.1). This type of navigation is also used in navigating set-top box menus. Similar to some automated teller machines, soft-key interfaces have two or more physical keys that are blank but are mapped to two or more options that are presented in the on-screen interface (Figure 7.1). The d-pad allows the user to move the cursor focus around the interface. Unlike computer navigation where a discrete cursor (mouse pointer) is moved around the screen almost as a proxy for the finger, the nature of interface movement on the mobile phone is more subtle (often a highlighting) and more constrained.

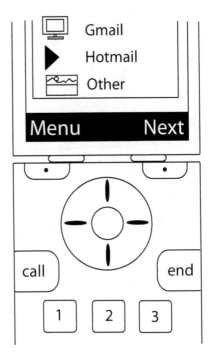

Figure 7.1 The top half shows a typical mobile phone interface with two soft-key options. The soft-key options (Menu, Next) are mapped to two physical keys located below them.

Movement is usually limited to active interface elements (between text boxes and buttons). Frequently, up/down movement is meant for within-page navigation, while left/right movement is used to navigate between pages or tabs.

Soft-key interfaces are space efficient because they allow multiple commands to be mapped onto a limited number of buttons. The commands change as the situation warrants. However, this benefit can also be a major source of usability errors. Mode errors, errors induced by the user not realizing which mode they are in, may be common, especially if a certain soft-key always represents a certain action (e.g., it is always the "Cancel" key but suddenly becomes a different function). In addition, the constantly changing relationships between button and function may be more difficult to learn for older adults.

Section 7.3.4 illustrates the task of composing an e-mail message. Some of the design issues discussed in the previous chapters will be illustrated using this example. As a disclaimer, the solutions we provide are meant only to illustrate one possible, general interface redesign solution. The following section is organized as follows: a user profile, task analysis of that user's task, and a list of the possible issues followed by recommendations.

7.1   Perceptual concerns

A major consequence of the small screen size is that interface elements are much smaller than they might be in another type of interface. Text, icons, and hardware buttons are all smaller than what users may be accustomed to. In addition, by virtue of their design, mobile phones may be more likely to be used outside the home in varying lighting and noise conditions. Screen displays may appear washed out and will be difficult to discern (Chapter 2). In addition, auditory feedback, if the phone provides any, may not be heard in the noise of the outside world (Chapter 3).

7.2   Cognitive concerns

User interaction with small-screened mobile devices presents many cognitive challenges. First, because of the small screen size, the interface cannot present as much information as a desktop computer application. Other than simple tasks such as dialing the phone, many tasks require multiple steps and therefore multiple screens of information, which places demands on attention and working memory. Another consequence of the small screen size is that phones often make use of hierarchically organized menu systems (where menu options are embedded in layers of categories). For example, changing the ringer sound may involve the user first selecting "System," then "Preferences," then "Sound," then "Ringers." From here, they may have to navigate a list of ringer sounds.

Other than touch-screen phones, which use a direct input device whereby the input device is embedded within the user interface, the user interface (display) and input device (number pad, QWERTY keyboard, or soft-keys) are separated by some distance. This arrangement may require the user to frequently switch their attention between the screen and input device. As discussed in Chapter 4, this increase in information access costs heightens the possibility of user error because the user's attention is on the input device and not on the display. Finally, the user interaction experience with mobile phone interfaces may not cleanly replicate user interactions with other types of consumer devices. For example, the multistepped nature coupled with the small screen size means that users may not easily be able to obtain a global picture of where they are in the task. This means that users will not be easily able to use their knowledge of other systems (such as the web, computers, or consumer electronics).

7.3   Usability assessment

The major issues that older adults are likely to have with mobile phone interface usage are centered on cognitive (information organization, screen and task layout) and perceptual issues (size, color, and location of interface elements) exacerbated by the very small screen size. To determine the type of problems that an older adult is likely to have and how they are likely to behave, we will start with a profile of a likely older user, their relevant characteristics, the tasks they are likely to want to accomplish, and a scenario or story that involves our user engaging in the tasks. Finally, the mobile phone tasks are presented in a task flow that shows what the user is likely to see, what actions they take, and what feedback they receive. Below each screen image is a revised design that attempts to alleviate the main issue.

7.3.1   User profile and persona (Figure 7.2)

- 65 years old
- Has worked in the public school system for 35 years; retired 3 years ago
- Has four children (all grown) that live out of state, and six grandchildren

Figure 7.2 Marilyn Smith, retired school system employee and grandmother.

- Spends her free time volunteering at the local food bank and managing their donations list

7.3.1.1 Technological experience

- Marilyn has a 4-year-old PC (with a 17-inch monitor) running Windows XP at home. She still uses dial-up internet.
- Her main computer-related tasks involve sending and receiving e-mails and occasionally looking up information about her hobbies (flower arranging and golf).
- She uses a web-based e-mail service with the Internet Explorer browser.
- She has had a cell phone for a few years, but recently received a more advanced feature-phone from her children with e-mail and internet capabilities. She has slowly learned to use short messaging service (SMS) and multimedia messaging service (MMS) functions to send pictures directly to her grandchildren's phones but her daughter does not use SMS (only communicates via e-mail).

7.3.2 Tasks

Marilyn has not yet fully explored the full capabilities of her phone, but being out of the house frequently, she would like to use the messaging and e-mail capabilities to communicate with her husband and her children and grandchildren. She is especially interested in showing off pictures sent to her by her grandchildren and sending pictures back to them. Specifically, she is interested in:

- Checking e-mail
- Viewing pictures
- Sending e-mail (and sometimes SMS)
- Looking up information on the web (usually starts with the default search engine)

7.3.3 Motivations, attitudes, and current problems

Marilyn is familiar with computers (having used them at her job). However, she has very basic experience with mobile technology and the internet. Her first phone was a basic mobile phone that could only make voice calls. Her attitude toward technology is utilitarian, and she expects things to work with as little fuss as possible. She thinks her new phone is interesting, and she is excited about being in regular contact with her grandchildren. However, she thinks that there are many features that she does not want or need, and she is bothered by how complicated seemingly simple tasks seem to be.

7.3.4   Scenario

Marilyn is on a day trip with friends to the local botanical gardens. She sees a picture of a flower species that she wants to know more about. Her daughter is a horticulturalist at a garden center, so Marilyn snaps a picture with her new phone and wants to send it to her daughter. She is able to take a picture relatively easily, but is now attempting to send it as an e-mail. Unfortunately, she has not yet set up her e-mail account on the phone. She starts by searching for the e-mail program icon. She is familiar with e-mail on a computer, but the interface on the phone is foreign to her.

Table 7.1 is an example task analysis and proposed redesign of Marilyn's phone display, specifically the task of sending an e-mail. Our

Table 7.1 Setting Up Mobile E-Mail

Step 1	The location of the OK menu option does not conform to convention of desktop user interfaces. It is also not clear how Marilyn is supposed to quit.
After accepting, the user is asked to give the application permission to access the network.	

Recommendations

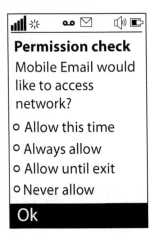

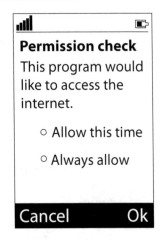

Figure 7.3a Original screen.

Figure 7.3b Redesigned screen.

Observations

This is the first dialog that Marylyn sees after selecting the e-mail program icon. The top line is littered with icons, many of which are unnecessary for the current task of e-mail.

It is not clear how the options of **Allow this time** and **Allow until exit** are different or clear why so many options are needed.

The wording and options have been simplified. Extra icons in the top status bar have been removed.

The location of the OK menu option should be located on the right to maintain consistency with desktop UI standards. A cancel option should be included, since the user is about to start a multistep process.

Table 7.1 Setting Up Mobile E-Mail (Continued)

Step 2	Recommendations

User enters their mobile telephone number.

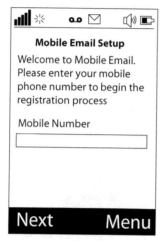

Figure 7.4a Original screen.

Observations

The next screen asks for the user's phone number, but Marilyn has not memorized her number. Out of frustration, she thinks about just not sending the message, but she finds a piece of paper that has her telephone number and enters it.

Not yet used to the idea of entering area codes, she forgets to enter the area code, which gives her an error message. The input field should chunk numbers by area code/exchange (e.g., 555-555-5555).

Figure 7.4b Redesigned screen.

The user now only has to confirm their telephone number.

Step 3

While the system processes the user's information, they see a feedback screen.

Figure 7.5a Feedback screen.

(continued)

Table 7.1 Setting Up Mobile E-Mail (Continued)

Observations	Step 4

Observations

After Marilyn enters her phone number, she receives a screen indicating that something is happening, but it is not clear what is happening or how long it will take.

If the wait time is longer than a few seconds, it should inform the user of progress.

Recommendations

Figure 7.5b Redesigned screen.

The feedback message identifies more clearly what is happening and provides a progress indicator so the user can exit the process if it takes too long.

Step 4

After processing, the user sees the screen to create an e-mail account.

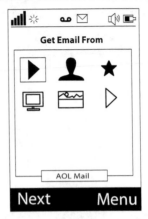

Figure 7.6a Original screen.

Observations

After waiting a few seconds, Marilyn is presented with a screen of icons. As she examines the icons she notices that there are no labels.

The icons are not clear in meaning. The text label for the icons is located on the lower edge of the interface.

Recommendations

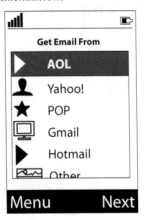

Figure 7.6b Redesigned screen.

Table 7.1 Setting Up Mobile E-Mail (Continued)

Move the icon label closer to the icon, or move the icon label to the "Get e-mail from" line (i.e., "Get e-mail from: AOL Mail").

AOL and next arrow icons are similar in shape, color, and size. Use a different arrow icon.

Instead of an icon gallery, the list has been organized into a list with the labels next to each icon.

Composing an E-Mail Message

Step 1	*Recommendations*
Within the mobile e-mail application, user selects "Menu" (right soft-key).	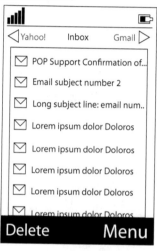

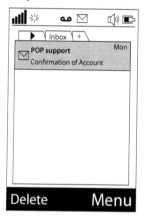

Figure 7.7a Original screen.

Figure 7.7b Redesigned screen.

Observations

It is not intuitive as to how to navigate to the different tabs (use of the left/right directional pad). The initial view is the inbox for one of the mail accounts. If other mail accounts are created, they show up as icons within alternate tabs. From this screen, Marilyn is not clear how to compose a new message, so she presses the Menu soft-key.

Instead of tabs, the new interface uses a carousel of left/right arrows. When the user clicks the right or left arrow, the dimmed option (in the screen image above either Yahoo or Gmail) takes central focus.

(continued)

Table 7.1 Setting Up Mobile E-Mail (Continued)

Step 2	Recommendations

Step 2

In the pop-up menu, user selects the "Write" function. User can use the directional pad or press the 5 key.

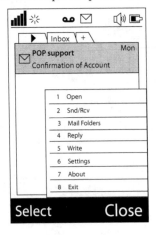

Figure 7.8a Original screen.

Recommendations

Figure 7.8b Compose a message.

The menu options have been rearranged to reflect possible frequency of use (more frequently accessed commands at top, less frequent at bottom).

Observations

After pressing the menu, Marilyn sees the "Write" option. However, its location in the middle of the menu is confusing. The options should be organized or grouped by similarity of function.

There are three ways a user can select an option from this menu: (1) by hitting the select soft-key, (2) hitting a specific number, (3) hitting ok/menu hard key. This seems redundant; remove the select soft-key.

Step 3

User enters common mail fields (TO, CC, Subject, Body). When finished composing e-mail, user selects "Send" (left soft-key).

Figure 7.9a Original screen.

Table 7.1 Setting Up Mobile E-Mail (Continued)

Observations	**Step 4**
From this point, Marilyn starts to enter the e-mail address of her daughter. However, she is not certain about the spelling. The address book is accessible from this screen, but it is hidden in the Menu soft-key. The space for the body of the message looks small and similar to the other fields.	User sees a feedback screen as the message is sent. When the message is successfully sent, user is taken back to their inbox.

Recommendations

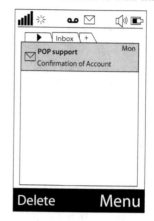

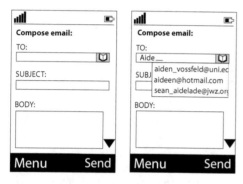

Figure 7.10a Original screen.

Figure 7.9b Redesigned screens.

Observations

It is recommended that frequently used e-mail addresses (or last used addresses) be memorized and appear when the user types in a few characters.

There is no feedback of the sent message.

The text field should also indicate that the address book is available.

Recommendations

First panel illustrates the new address book icon (accessible with a right keypress). Second panel shows autocomplete when user starts typing. The system will match to nearest address book entry. OR the user can hit right soft-key to put focus on the book icon (within the field) and click to see the last-used e-mail addresses.

Figure 7.10b Redesigned screen.

(continued)

Table 7.1 Setting Up Mobile E-Mail (Continued)

The system should offer a brief message (possibly a modal box that times-out or an auditory or tactile [vibration] feedback.
A modal dialog gives feedback after a sent message. The user sees this dialog after the first sent message but can turn it off for subsequent messages.

recommendations focus on the needs of older users. Each step in the task flow for the analysis is named and contains an illustration of the navigation system, followed by potential usability issues, our recommendations, and a redesign of that step in the display.

7.4 Specific design changes/recommendations

The overall problem stems from presenting too cluttered an interface for the tasks that the user wants to accomplish. This is exemplified by the cluttered status bar at the top of the screen that presents the status of the battery, network coverage, voice mail, e-mail status, and volume. Although the top section represents a convenient place to show such global status information, it adds to visual clutter. The main issues are summarized in the next sections.

7.4.1 Perceptual

- Size: The size of many user interface elements is extremely small. This is primarily a consequence of the small display size.
- Readability: One consequence of the physically small display size is that there are relatively few pixels available. The result is a blocky display with pixels that are apparent. These types of displays (with a low pixel density) are not good candidates for font smoothing technologies.
  - In addition, for the majority of phone handsets, font readability is low. A default sans serif font is often used that is not specially designed for readability on small screens (highly condensed, tight line spacing).
- Color cues: Although not readily visible in the screen captures, different colors are used to indicate focus/selection at different points in the task. Sometimes gray is used to indicate focus, while at other times yellow or green is used.

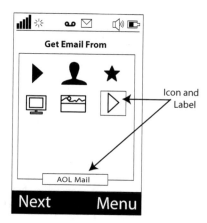

Figure 7.11 Icon and labels are not in close proximity.

- Recommendation: Consistency is a key aspect of usability. Color is an important cue for users and should be as consistent as possible throughout the interface.
- Blue colors are used for some text. Older adults are less sensitive to varying shades of blue.
- Contrast for certain key user interface elements is low. For example, using yellow on light gray in a tab.
- At least in Western cultures, the visual flow of an interface should proceed from upper left to lower right (same as reading a page). Decision actions (like NEXT or DONE) should therefore be located in the lower right to be consistent with expectations.
- Icons and labels are located very far away from each other. For example, in Figure 7.11 (which represents different types of e-mail accounts), the label for each icon is located on the bottom edge of the screen (e.g., AOL mail).

7.4.2   *Cognition/attention/memory*

- Clutter: The display presents much more information than is necessary for the task, leading to clutter which demands attention (e.g., the status bar icons).
- Position: The interface is consistent in the placement of the MENU option (which is always in the lower right). However, mobile phone interfaces can easily change the function assigned to that location. For example, that position/key sometimes functions as CLOSE or BACK.

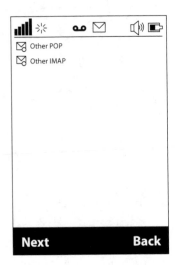

Figure 7.12 In this screen, the left side is mapped to the NEXT function, while the right side is mapped to the BACK function, which is opposite most expectations.

- Interface consistency is a general usability issue. However, it may be even more important for older adults. Cognitive aging research shows that inconsistency (i.e., colors meaning different things or functions changing positions) has greater detrimental effects for older adults. These effects include slower performance and longer time to learn.
- Icon consistency: The use of icons is not consistent with function. Similar icons are used for different screens and look too similar.
- Mapping: Mobile phones use soft-keys for functions (the lower left and right areas of the screen). These areas should be consistently mapped to conceptually similar functions. One natural mapping (in cultures that read from left to right) is that values increase from left to right (left representing negative or back and right representing positive or forward). Figure 7.12 and Figure 7.13 are an example where these mappings are not consistent.
  - One reason for the inconsistency is that some options are designed for easy one-handed use, and placing the NEXT option on the right would be difficult to reach (for a right-handed user).
- Indication of steps: There is no notification of where the user is in the task (what step).
- Accelerator keys are possible (numbers mapped to functions), but the number mappings change (compose may be 5 on one screen but 6 on another).

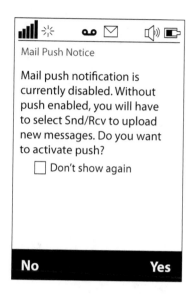

Figure 7.13 However, on this screen, the LEFT is mapped to NO (negative), while the right side is mapped to YES (positive), which is consistent with most Western user's expectations.

7.5   Summary

The approach taken in the redesign of the mobile phone interface typifies the statement that usability can be achieved through many changes that may seem trivial on their own. However, when combined, they can lead to measurable increases in user performance and satisfaction.

7.6   Suggested readings and references

Ryu, Y. S., Babski-Reeves, K., Smith-Jackson, T. L., and Nussbaum, M. A. (2007). Decision models for comparative usability evaluation of mobile phones using the mobile phone usability questionnaire. *Journal of Usability Studies, 3*(1), 24–39.

Valdez, A. C., Ziefle, M., Horstmann, A., Herding, D., and Schroeder, U. (2009). Effects of aging and domain knowledge on usability in small screen devices for diabetes patients. *Lecture Notes in Computer Science, 5889/2009,* 366–386.

Ziefle, M. (2002). The influence of user expertise and phone complexity on performance, ease of use, and learnability of different mobile phones. *Behaviour & Information Technology, 21,* 303–311.

Ziefle, M. (2008). Instruction formats and navigation aids in mobile devices. *Lecture Notes in Computer Science, 5298/2008,* 339–358.

chapter eight

Integrative example
Set-top box

One display that is or will soon become common is that of the set-top box. More than half of the U.S. population subscribes to a cable television service (National Cable & Telecommunications Association). Among those users, 61% subscribe to digital cable service. Digital cable promises a better quality picture and more channels, but users must typically use a set-top box to receive these benefits. A set-top box (STB) is a device that provides the user a graphical user interface to navigate their channels using a remote control and, in some cases, control advanced services such as television recording.

STB interfaces can be as intricate as an interface in a desktop computer application. A further complication is that navigation is achieved with a remote control device, often with directional arrows and numbers. In the conceptual redesign of STBs, we will focus on the on-screen interface and display. Similar to the other examples in this section, potential older-adult issues are categorized in terms of perceptual (seeing and hearing) and cognitive (thinking and remembering).

The STB interface is a graphical interface, dominated by tables of information, forms, and buttons for selection. Selection is done solely using a remote control device and using directional arrows. There is some capability to use accelerator keys (e.g., numbers) to reach specific parts of the interface (Figure 8.1).

Because of the highly constrained nature of interaction (using a remote control), the design of the interface primarily relies on focus box movement, selection of buttons, checkboxes, lists of items, and limited text input. There is also an assumption that users will know that only a certain subset of the remote control buttons are appropriate for certain portions of the interface. This is a kind of mode error, or an error induced by the system not clearly indicating what mode it is currently in. For example, when in the guide view, the primary navigation buttons are the directional arrows and the central selector button ("enter"). If a user inadvertently presses a special button (for example to connect to the Internet to order more channels) he will immediately be taken out of their current task flow without any indication of why or how. These mode changes can be jarring and frustrating for the user with

Figure 8.1 Set-top box interface and remote control.

an unsteady hand who accidently clicked a button. An overview of the process of setting a timer to record a program in the future is outlined in Table 8.1.

8.1 Cognitive concerns

The STB is a graphical and menu driven interface. The interface is used to navigate the content (recorded shows, television listings for up to 2 weeks). Because of the limited user interface possibilities and canvas size (usually, interface elements are large to accommodate sofa-distance viewing) many options are nested in other screens or menus. This would tend to increase the working memory burden and also reasoning burden as users have to "figure out" what the next step is (especially if a common metaphor is not used).

8.2   Perceptual concerns

STB interfaces are typically used on televisions, so we can assume that the user will be some distance away. Depending on the size of the television and the viewer's distance, items can be difficult to perceive.

8.3   Usability assessment

8.3.1   User profile and persona

- 67-year-old male (Figure 8.2).
- Worked as an engineer for a major aerospace company for 30 years.
- Is active in the community and an avid golfer. Competes in senior golf tournaments.
- Don represents a typical healthy 67-year old. He has high blood pressure and takes medications to control his cholesterol.

8.3.1.1   Technological experience

- Is very comfortable with some technology (computers), having used them at work. He is mainly familiar with Windows XP.
- Owns a basic cell phone that makes calls (no internet) but only uses for emergencies.
- Is comfortable with technology and willing to invest some time in learning to use things that will help him with his tasks. However, he does not seek out novelty and will rapidly lose patience with tasks that take too many steps or are ambiguous.

Figure 8.2 Don Jones, retired engineer. (From www.sxc.hu/photo/617485.)

8.3.2   Tasks

Being out of the house frequently, Don does not have a chance to see many of his favorite television shows (as well as the local news). Don recently upgraded his television to high definition (HD), which necessitated purchasing a new STB with time-shifting capabilities. Previously he had used analog cable, which did not require a box. His primary task is recording the local news as well as golf programming. He is familiar with VCR technology, but is having some difficulty understanding the options available with digital video recording (DVR) offered with his STB.

The task flow to record a program with our STB can seem un-intuitive because so many options are presented, all on one screen without any orientation provided to the user as to where to start and end. In this case, a more guided, wizard-style interface might be more appropriate. Another way to handle the task complexity is to dramatically reduce the number of possible actions available to the user (but keep them available). This approach was taken in the current redesign. Our user is likely to only set up recordings and to do this as quickly and as easily as possible.

8.3.3   Current problems

- Understanding recurring recordings
- Setting one-time recordings (special events)

8.3.4   Scenario

Don is going on a cruise along the Mediterranean that will last 2 weeks. However, during that time the Masters Golf Tournament will be on television. He would like to record one of the days of the event on his DVR. He has had the STB for several months and has only tried some of the features (pausing live television, rewinding). He has not yet utilized the scheduled recording functions.

Table 8.1 is an example task analysis and proposed redesign of a STB display, concentrating on the task of recording a program. As with all task analyses and redesigns, we focused on the needs of older users when we created the recommendations and redesign for each step in the recording process. Each step in the task flow for the analysis is named and contains an illustration of the STB display. Our observations regarding that step in the task flow and display follow, pointing out potential usability issues. Finally, we present a redesigned display screen and a list of recommendations that were taken into account when improving the display for older adults. As a disclaimer, the redesigns we provide are meant only to illustrate one possible solution.

Table 8.1 Task Flow: Record a Program (Once)

Step 1

Press the **"guide"** button on remote control, and find desired program, and then click the **SELECT** button on remote control (or the record button)

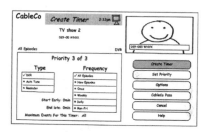

Figure 8.3a Original screen.

Observations

- Clicking the record button (or select) brings up a task priority screen (next panel).
- The interface also assumes (and encourages) multitasking by showing a live window of the current show in the upper right. Some users may be distracted by it. The live view should be moved to a less important area of the screen.

Recommendations

Figure 8.3b Redesigned screen.

- When a user clicks the record button in the program guide, the single instance will record and feedback will be indicated on the guide (a red dot next to program name). To change additional parameters, they can click the record button again to bring up the next screen.

  - Single press RECORD = record once (indicated with a red circle on the listing).
  - Press RECORD again = record every time (indicated with multiple red circles).
  - Press RECORD again = turn off recording.
  - Press SELECT = setting screen (next panel).

Step 2

Press the **SELECT** button once the desired program is highlighted. The Create Timer screen appears.

Figure 8.4 Original screen.

(*continued*)

Table 8.1 Task Flow: Record a Program (Once) (Continued)

Observations	**Step 3**

Observations (left column)

- This should not be the first screen the user sees when they press the record button. If the user would like to just record this program, they can click "Create Timer." The focus starts on "create timer," but the user must navigate left to adjust the different settings. If the user chose the default values (DVR and All Episodes), they would inadvertently create a recurring timer.
- The user may be inclined to simply press the record button on the remote control. However, pressing the record button (or following this task flow) leads to similar steps. The user is presented with a relatively complex configuration screen given the simplicity of the task ("record a program").
- It is not clear what the "Priority" label indicates.
- Because there are so many discrete options on separate screens, each with a separate button on the left, it can cause confusion (especially because they are not sequential steps). The user is likely to get lost in the menu screens.

Recommendations

Removed screen
- Figure 8.4 is now an optional screen (only reached when the record button is double-tapped).
- The CableCo pass option is moved from its own button to the "recording type" because it is a special type of recording.

Step 3

Select recording type (DVR, auto tune, or reminder).

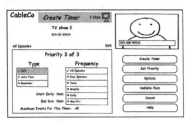

Figure 8.5a Original screen.

Observations

- The next actions are not clear. They may not know what each of the options mean.
- The available options are unclear (DVR, Auto tune, and reminder). DVR means to "record," auto tune will change the channel when the program starts (but not record), and reminder will display an on-screen reminder.
- Status information shows how much "padding" will be recorded with the show (to accommodate programs that start early or end late). These options are not changeable from this screen. The user is left to discover that they are only changeable from the "options" button.

Recommendations

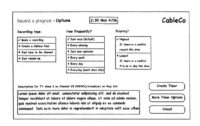

Figure 8.5b Redesigned screen.

Table 8.1 Task Flow: Record a Program (Once) (Continued)

Screens have been replaced by the single optional screen with options that are most likely to be changed (this information could have come from user research, observations, or usage logs). Less frequently changed options have been relegated to another screen (under "More timer options"). In addition, the number of buttons has been reduced.

Step 4

Set the frequency of recording.

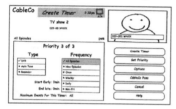

Figure 8.6 Original screen.

Observations

• The default option is to record all occurrences of that show ("all episodes"). However, the user has the option to change the recording frequency. Don's desired option (once) is third on the list.

Step 5

Optionally, the priority level of the program can be set by selecting the "Set Priority" button.

Figure 8.7a Original screen.

Observations

• If there are multiple recordings in the queue, and they conflict (they are scheduled to record a program at the same time but on different channels), setting the priority determines which program overrides.

Recommendations

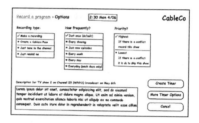

Figure 8.7b Redesigned screen.

• This step has been replaced by a single option screen.

Step 6

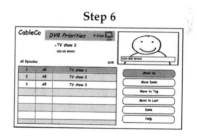

Figure 8.8 Original screen.

Observations

• In the priorities screen, the system shows all scheduled recordings and their priorities. The user can move specific shows up or down the priority list by using the buttons MOVE UP or MOVE DOWN.
• When in this screen, it is not clear where the user is currently located in menu hierarchy.

(continued)

Table 8.1 Task Flow: Record a Program (Once) (Continued)

Recommendations

- This level of control may not be useful to the user, and even presenting it as an option merely adds clutter and may contribute to confusion about the steps involved in the task.

Step 7

After setting the priority, the user clicks Done and returns to the Create Timer screen. From there, they can configure Options.

Figure 8.9 Original screen.

Observations

- The plethora of buttons leads to confusion.

Recommendations

Step 8

From the Timer Options screen, the user can configure how early or late the timer starts.

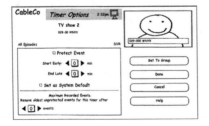

Figure 8.10a Original screen.

Observations

- The exact meaning of start early/end late is confusing.

Recommendations

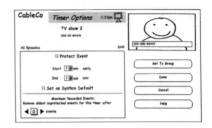

Figure 8.10b Redesigned screen.

- The left/right arrows to indicate minutes were changed to an up/down orientation and embedded within the statement "start..." and "end..." to be more intuitive as to what action the user was doing.

Step 9

Returning to the Create Timer screen, the user can finally elect to create a recurring timer (in this case called a "pass"). User clicks "CableCo Pass."

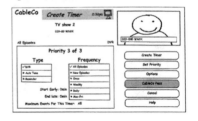

Figure 8.11 Original screen.

Recommendations

Table 8.1 Task Flow: Record a Program (Once) (Continued)

Step 10

From the Pass screen, the user can type in the name (program title, star, etc.). When a program matching that text appears, it will automatically be recorded. Optionally, there are two suboptions (set priority and options) that are configurable at this screen.

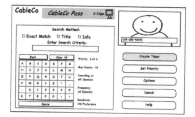

Figure 8.12a Original screen.

Figure 8.12b Redesigned screen.

- Instead of opening up a new screen, a modal dialog appears on top of the interface. The modal box allows the user to retain their task orientation (know where they are).

Step 11

After the options have been selected, the user can then create a timer. (by clicking "Create Timer").

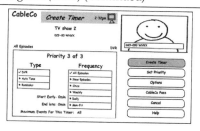

Figure 8.13 Original screen.

Observations

Recommendations

[already addressed]

Step 12

Feedback is given in the program guide. The show has a green symbol.

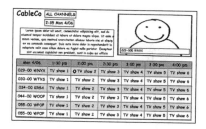

Figure 8.14a Original screen.

Observations

Recommendations

Figure 8.14b Redesigned screen.

8.4   Specific design changes/recommendations

8.4.1   Perceptual

- Sound: Sound is not used to indicate actions (e.g., button presses). This may be problematic for this type of interface because of the slight lag between action and interface reaction due to the wireless remote. Not having feedback sound could lead to unintentional double-clicks.
- Readability: Most displays (especially the television guide) are dense with text. Readability may be enhanced through the use of grouping using colors (show titles are white, descriptions are gray).

8.4.2   Cognition/attention/memory

- Clutter: One consequence of the sheer number of actionable items (e.g., buttons) is that next actions are not obvious. Clutter was reduced by moving options to an optional screen.
- The user/system dialog is not natural or simple. The simple task of recording a program requires too many steps and too many screens and offers too many possible configuration options. The task was simplified by reducing the number of steps.
- Indication of steps: There is no notification of where the user is in the task (what step). This information should be conveyed using either text or a graphical progress indicator.
- Many screens have a hidden and an unconfigurable time-out period. If a user remains on any screen too long, it will kick them out back to the live TV view. The interval may be very generous, but it may be completely unclear to users how much time they have—introducing a time pressure (in addition to the time pressure experienced by users trying to set this before their show comes on or trying to leave the house). This may detrimentally affect older adults who are examining or reading a screen.

8.5   Summary

The main changes with the STB revolve around a dramatic simplification of the task by moving configurations to a second level only accessible to those who want to change them. For the everyday task of setting a recording, the task is now as simple as pressing the record button. Additional advanced options have been "cascaded" such that options more likely to be changed are presented in a unified screen. Options that are even less likely to be altered (e.g., extending or shortening the default record time) are relegated to an even deeper level menu.

8.6   Suggested readings

Eronen, L. and Vuorimaa, P. (2000). User interfaces for digital television: A navigator case study. *Proceedings of the Working Conference on Advanced Visual Interfaces.* Palermo, Italy: ACM.

Pan, Y. and Ryu, Y. S. (2009). Insights for the TV interface from the mobile phone interface. *Journal of Usability Studies, 4*(4), 1166–177.

chapter nine

Integrative example
Home medical device

Many older adults have home devices with which they maintain their health and independence. Some examples include blood pressure meters, pulse oximeters, heart rate monitors, and blood glucose meters for persons with diabetes. We use the blood glucose meter (BGM) as an example of a seemingly simple device that can pose unintended problems for older users. It is important to keep in mind that the purpose of this chapter is not to problem-solve solutions for a single brand or product, but to illustrate the types of issues to look for on any device. The BGM is a good example, because past research has shown that older patients have difficulty with these devices and with monitoring blood glucose in general.

No matter what meter is used, the task of monitoring diet and exercise is complex. A participant in a study by Helen Klein and Amy Meininger exemplified this difficulty with the comment: "I wondered why my blood sugar was so high. Then I remembered that Big Mac. I guess that wasn't a good idea" (p. 722). Other work pointed out that even a "simple" task with a meter required over 50 steps.

Unfortunately, solving a problem with a particular BGM is not helpful for future designs unless it gives designers a theoretical understanding of what adversely affects the performance of older adults. The meter shown in this chapter could be easily improved, but the take-home message is that the future design of *all* devices will benefit from an understanding of the cognitive, perceptual, and motor concerns of older users. As with many systems, BGMs can contain more features than necessary or desired. This chapter provides a conceptual redesign of one system, specifically for older novice users. The technology used by BGMs may progress, but the principles behind a good interface for aging users will remain the same.

9.1 Cognitive concerns

As discussed, managing diabetes is a complex task before a glucometer is involved. Blood sugar is affected by carbohydrate intake, exercise, insulin, and even time of day. The design of the glucometer needs to support an understanding of these variables to help the user best manage blood sugar levels.

Some specific cognitive concerns are the number of steps a meter requires to achieve the current goal, how many advanced features are available (and how many are necessary to support the task).

Other cognitive concerns are the affordances of the device and materials. Do they meet the expectations of the users? Do they take advantage of prior knowledge?

Last, does the meter provide useful information? Literally, can the user understand and use the information as it is provided? Is there extraneous information present, or is the wrong information emphasized? Distractions and the need to inhibit distraction will absorb many of the resources needed for the already complex task.

9.2   Perceptual concerns

It is difficult to balance the need for portability with the need for visibility. It is also difficult to balance the need for low cost with visibility. All important information (the only information that should be displayed) should be of at least 14-point font. Buttons and other inputs act on the display and must also be perceived. For example, in a study done by one of the authors, it was noticed that novice BGM users had difficulty locating the "C" button, which was on the top of the device rather than the front. Even before an analysis of whether it should be called the "C" button, it must first be perceptible to the user as a button. Blinking text or information also make the display more difficult to read.

9.3   Movement control and input

As with visibility, the small size of portable devices can make operation difficult for those with motor control issues. Buttons should have surfaces with friction, not glossy surfaces, and provide feedback when pressed. Buttons should not be located where the device is typically gripped. Target areas of any kind, including test strips, should be large and stable.

9.4   Usability assessment

Although even expert users have difficulty with infrequent tasks on home medical devices, include older novice users in the usability testing process. Assign tasks beyond testing of their own blood, especially crucial but infrequent tasks. Test in environments cognitively similar to the ones where they will use the device. For example, these devices are mobile, and users will need to know what to do when the instruction booklet is at home. Test with and without the instruction booklet.

Figure 9.1 Paul McAllen: part-time business consultant; avid gardener; diagnosed at age 71 with Type II diabetes.

9.4.1   User profile and persona (Figure 9.1)

For an analysis of a BGM, user information would be gathered from a number of sources including persons with diabetes, older adults recently diagnosed with diabetes, and health-care professionals who work with these populations. Data could be collected via surveys and interviews to determine the most common tasks, the most difficult tasks, and the kinds of mistakes users of the BGM tend to make. These data could be augmented with results from the research literature where it is possible to discover common errors and misuses of this device. Once amassed, these data can be used to create personas that direct the thinking of designers during the design or redesign process.

- 72 years old.
- Part-time business consultant.
- An avid gardener.
- Paul was diagnosed with Type II diabetes 6 months ago. He was given a BGM to keep track of his blood sugar levels.
- Paul is having trouble changing his diet to match the prescription of his doctor.

9.4.1.1   Technological experience
- Paul works 20 hours a week, uses a computer for work daily.
- Paul considers himself comfortable with new technology and uses a cell phone and computer almost daily.
- He admits that he sticks with the basic features of the phone and does not check his voice mail.

- On the computer, he uses spreadsheet and word processing programs daily.
- He has activated some of the accessibility features of the software, mostly increased size so he can sit further back from the computer.
- Paul uses web access for e-mail only.

9.4.2   Tasks

Paul tests his blood sugar once every few days. He would like to avoid this task as much as possible, but he is also afraid of the consequences of ignoring his condition. He would like to do his testing discreetly, quickly, and be able to use the results with minimal translation. For example, he understands that a blood sugar level of 100 is optimal, but does not understand the seriousness of the numbers. He wonders if a level of 120 is "twice" as bad as a level of 110, but does not know where to look for this information. Paul is also interested in keeping track of his levels over time because his doctor has requested this information from Paul. Paul would not say he is interested in checking the calibration of his meter, but this is due to not understanding how the meter can function incorrectly when miscalibrated. Paul believes that setting the number on the meter to match his test strips is "calibration" of the meter and does not know that true calibration must be done with unexpired control solution.

9.4.3   Motivations, attitudes, and current problems

Paul says he feels comfortable with technology, but is frustrated by his BGM. He has only "calibrated" it twice since his diagnosis 6 months ago, despite the instruction to do so once a week, or if the meter is dropped, or if it gives strange readings. Because of this, he does not know that his meter is currently overestimating his glucose levels by about 10%.

Also, Paul believes he has always had a healthy diet and is having trouble forcing himself to count carbohydrates as directed by his doctor. He has become used to feeling a little ill most of the time due to the diabetes, and that makes him have more trouble recognizing when he should test his blood sugar.

9.4.4   Scenario and task analysis

Paul has had a hard time accepting his diabetes diagnosis. Despite his doctor's admonitions, he sometimes grabs food when passing through the kitchen and does not record the carbohydrates it contains. This happens most frequently during the morning, when he is working on the computer. He has also taken a few of the rules made by his doctor and created his own, sometimes dangerous, heuristics. For example, sometimes when his blood sugar tests high, he will go without eating to lower it rather than administer insulin.

Today, he feels "off" and wants to know if it is related to his sugar levels. He is sitting in his parked car, about to head into the gym. He decides to do a blood sugar test and gets out his kit that contains the meter, test strips, and needle. The day is very bright, and the glare and light wash out the screen on the meter. Paul has a suspicion his meter needs recalibration, but cannot really remember how to calibrate it and is wondering whether it is worth the fuss.

Table 9.1 is an example task analysis and proposed redesign of a sample BGM display and controls. The task analysis contains observations of difficulties Paul may experience and recommendations on removing those difficulties through a redesign. As with all task analyses and redesigns, we concentrated on the needs of older users in the recommendations section for each step. Each step in the analysis is named and contains an illustration of the BGM at that point in the task flow. Our observations regarding that step in the task flow and display screen follow, pointing out potential usability issues. We follow our observations with a redesigned display and the recommendations that were taken into account when redesigning the display. The redesigns we provide are meant only to illustrate one possible solution.

Table 9.1 Task Flow: Control Solution Test for Blood Glucose Meter

Step 1

Meter must be off.

Figure 9.2a Task flow: control solution test for blood glucose meter.

Observations

- Although the first step is always to make certain the meter is off, the off button is neither intuitive nor visible.
- The location of the off button, near the edge of the device where it is gripped, makes it prone to accidental activation errors.

- The arrow buttons are used for many purposes, including turning the meter on and off, scrolling through results, and other inputs. This invites mode errors, particularly when the different modes are not easily visible. It is difficult to tell what "state" the meter is in and thus what effect the arrow buttons will have.
- All text except numbers is less than 12-pt font.
- mg/dL is in a small font, and the information mg/dL provides is not helpful to the person with diabetes.

(*continued*)

Table 9.1 Task Flow: Control Solution Test for Blood Glucose Meter (Continued)

Recommendations

Figure 9.2b Revision of Figure 9.2a.

- Provide on/off button.
- Arrow buttons used only for increasing or decreasing values and moving through wizard-like interface.
- Remove nontask information.
- Enlarge text, screen, and increase contrast of screen.

Step 2

Check the code on the test strip vial before inserting the strip.

Figure 9.3 Task flow: control solution test for blood glucose meter.

Observations

- Needing to check both vial and meter violates the proximity compatibility principle.
- The container must be used in tandem with device, though only the strips interact with the device.

(No picture)

Recommendations

- Automate calibration of strip and meter.

Step 3

Make sure the three contact bars on the test strip are facing you.

Figure 9.4a Test strip.

Observations

- The strip can fit into device with contact bars facing the wrong way.
- Noticing the contact "bars" requires high visual contrast sensitivity.
- The correct position for the strip appears to be upside down, violating expectancy.

Recommendations

Figure 9.4b Redesigned test strip.

- Create affordances in strip and device so strip will only fit correctly into meter.
- Mark or label "top" of strip.

Table 9.1 Task Flow: Control Solution Test for Blood Glucose Meter (Continued)

Step 4

Do not bend the strip. Push the strip in
as far as it will go.

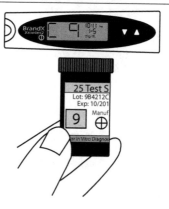

Figure 9.5 Inserting the test strip into the
meter.

Figure 9.6a Comparison of codes.

Observations

- Fine motor control is required to fit
  the test strip into the opening without
  bending the strip.
- Accuracy in this task is aided by port
  shape.

(No picture)

Observations

- Mode error potential: The code
  buttons are also on/off buttons for
  the meter.
- The contrast on screen is low
  (charcoal on gray).

Recommendations

- Supply sturdier strips.

Recommendations

Step 5

If the code on the meter does not match
the code on the test strip vial, press ▲
or ▼ to match the code number on the
test strip vial.

Figure 9.6b Redesigned screen.

- Increase contrast ratio on screen.

(continued)

Table 9.1 Task Flow: Control Solution Test for Blood Glucose Meter (Continued)

Step 6	Step 7

The new code number will flash on the display for three seconds, and then stay constant for three seconds. The display will advance to the screen with the flashing blood drop icon.

Figure 9.7a Indication the meter is ready to receive a sample.

Observations

- Remembering the code requires short term memory.
- Flashing of the code is difficult for those with poor temporal resolution and visual acuity.
- Drop icon indicates readiness, but is misleading. Other steps must occur before solution applied.

Recommendations

Figure 9.7b Redesigned screen.

- Code remains on screen.
- Environmental support from arrow to indicate how to progress.

If the codes already match, wait three seconds.

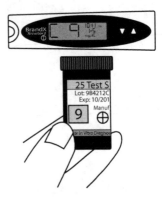

Figure 9.8 Code comparison.

Observations

- Instructions to user contain a conditional statement that requires working memory.

Recommendations

(No picture)

Step 8

The display will advance to the screen with the flashing blood drop icon.

Figure 9.9a Drop icon screen.

Table 9.1 Task Flow: Control Solution Test for Blood Glucose Meter (Continued)

Observations

- Icon meaning needs testing with older adults of different vision levels. It could appear to be many other symbols, from a candle to a landscape indicating time of day.
- Fine lines difficult to discern and require high visual acuity.
- The view of the strip on the screen is not matched to view of the strip by the user while viewing the screen.
- The "mg/dL" label has no meaning on this screen.

Recommendations

Figure 9.9b Redesigned screen.

- Replace icon with instructive text.
- Remove fine lines.
- Remove unnecessary information.

Step 9

Press ▲ so that the control solution test symbol CtL appears in the upper right corner of the display.

Figure 9.10a CtL solution screen.

Observations

- Combination of capital and lowercase strange, and occurs through choice of a screen with limited alphabetic options.

- mg/dL clutters this screen and step.
- Mode errors for same control now used for symbol change.
- The fine lines are difficult to read.
- Spacing of letters strange, and again due to the screen choice.

Recommendations

Figure 9.10b Redesigned screen.

- Provide this information about being in the control test "mode" visibly, by putting Ctrl before the test result.
- Remove screen clutter, replace icons with large text.

Step 10

If you decide not to do a control solution test, press ▲ again to remove CtL from the display.

Figure 9.11a CtL removed.

Observations

- Mode errors for same control.
- mg/dL is clutter.

Recommendations

Figure 9.11b Redesigned screen.

(*continued*)

Table 9.1 Task Flow: Control Solution Test for Blood Glucose Meter (Continued)

- Once marked as a control test about to occur, the meter starts providing environmental support through a wizard-like interface.
- Arrow on interface signals use of arrow on keypad to progress to next step.

Step 11

Shake the control solution vial before each test. Remove the cap.

Observations

- This step not prompted by device and requires prospective memory from the user.

Recommendations

Figure 9.12 Redesigned screen.

- Wizard now prompts next step.
- Arrow on interface signals use of arrow on keypad to progress to next step.

Step 12

Hold the vial upside down and gently squeeze a hanging drop.

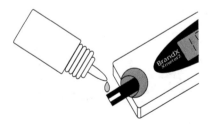

Figure 9.13a Application of control solution.

Observations

- This step is not prompted by device.

Recommendations

Figure 9.13b Redesigned screen.

- Arrow on interface signals use of arrow on keypad to progress to next step.

Step 13

Touch and hold the hanging drop of control solution to the narrow channel in the top edge of the test strip. Make sure the confirmation window fills completely.

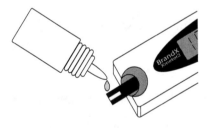

Figure 9.14a Application of control solution.

Observations

- The confirmation "window" is actually part of the test strip. It is low contrast compared to the strip itself, especially when liquid is applied.
- Blood reduces contrast.

Table 9.1 Task Flow: Control Solution Test for Blood Glucose Meter (Continued)

- The target area for the drop is small, and the motor skills required include estimation of how much to grip and squeeze the bottle with one hand while using both hands to direct the nose of the bottle to the strip without actually touching the strip to the bottle.

Recommendations

Figure 9.14b Redesigned strip.

- Application area (confirmation window) enlarged and of high contrast.

Step 14

When the meter detects control solution in the test strip, it begins to count down from 5 to 1.

Figure 9.15a Original count-down screen.

Observations

Recommendations

Figure 9.15b Redesigned count-down screen.

- No recommendations.

Step 15

Your result will then appear on the display, along with CtL and the unit of measure.

Figure 9.16a Original results screen.

Observations

- Here, time is labeled as mg/dL, because the mg/dL stays fixed on the screen.
- CtL looks like CEL due to the mixture of letters and the proximity of the "t" to the top of the screen.
- A control test is driven by the range printed on the test strip bottle and is not matched to human levels.

Recommendations

Figure 9.16b Redesigned results screen.

- Ctrl precedes the information it labels and is readable.
- Date and time are visible, because people often need to record their results with that information.
- 100 is always "perfect," the number for desirable blood sugar.

(continued)

Table 9.1 Task Flow: Control Solution Test for Blood Glucose Meter (Continued)

Step 16	Step 17
Compare the result displayed on the meter to the control solution range printed on the test strip vial.	If the results you get are not within this range, the meter and strips may not be working properly. Repeat the control solution test.

Figure 9.17a Original result comparison task.

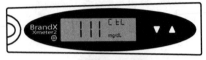

Figure 9.18a Task flow: control solution test for blood glucose meter.

Observations

- Each vial of test strips may have a different control solution range.
- Constant changing of what constitutes "correct."

Recommendations

- 100 is always "perfect," the number for desirable blood sugar.

Observations

- Identification of problem on this screen and a potential solution.

Recommendations

- 100 is always "perfect," the number for desirable blood sugar.

9.5 Specific design changes/recommendations

Perceptual

- Choose as high-contrast a display as possible.
- Choose as veridical a display as possible (avoid trying to make letters from number-based screens).
- Measure visual angle of all display components.
  - Increase perceptibility.

Cognitive
  - Remove unnecessary information, such as the "mg/dL" text.
  - Support user goals; do not add tasks.
  - Have as few hierarchical steps as possible.
  - Only show information required for the user goal.
  - Simplify the feature set, or hide advanced features.
  - Decrease complexity
    - Through shallow menus
    - By displaying choices
    - With environmental support
    - Through visual enhancement

Movement
  - Provide tactile and haptic feedback from input.
  - Choose a location for input that does not promote accidental activation.
  - Increase the size of targets for input. This includes buttons and the test trip.

9.6   Summary

For this redesign, we chose a relatively common home medical device for older adults. The usability steps included developing a user profile, creating a persona from that profile, analyzing a common task via task analysis, and redesigning the device based on the steps of that analysis. The redesign was informed by the areas covered by the first sections of this book: vision, cognition, hearing, and movement. Knowledge of these areas is critical for all the redesigns included in this book. Optimally, this redesign would be tested with older users before deployment. Keep in mind that these redesigns are conceptual, to be used for content and process ideas, and have not been tested on users.

9.7   Suggested readings

Klein, H. A. and Meininger, A. R. (2004). Self management of medication and diabetes: Cognitive control. *IEEE Transactions on Systems, Man, and Cybernetics—Part A: Systems and Humans, 36*(6), 718–725.

Rogers, W. A., Mykityshyn, A. L., Campbell, R. H., and Fisk, A. D. (2001). Analysis of a "simple" medical device. *Ergonomics in Design, 9,* 6–14.

chapter ten

Integrative example
Automobile displays

Although car navigation systems are becoming ubiquitous, especially for the rental market, they can offer more features than usability. This chapter provides a conceptual redesign of one system, specifically with older novice users in mind. The technology of navigation systems may progress, but the principles behind a good interface for aging users will remain the same.

We present an overview of the issues in each of the areas covered by the first sections of this book. Then, we present a task analysis of the current system followed by a redesign of the system for an older user. Keep in mind that these redesigns are conceptual and intended to be used for content and process idea but these designs have not been tested on users.

10.1   Cognitive concerns

Some of the cognitive concerns specific to a navigation system are the need for spatial ability to perceive and follow different map views, the need for working memory to manage menu navigation, and the need for inhibition of attention to the system when a driving task takes priority over navigation system directions.

Spatial ability, or the ability to manipulate the orientation of mental images, is required both by paper maps and navigation systems. There are three views often used by map and navigation systems: allocentric, egocentric, and a hybrid of allocentric and egocentric. An allocentric view means the user can see a representation of far beyond the actual field of view, as in a map of an area. The egocentric view is a first person view that includes what the user can see at any point. The hybrid view is a "hover" view, where more is visible than a first person view, but still limited by the direction and position of the user. Paper maps are universally allocentric, meaning that the system of roads and landmarks is symbolically laid out from above, without a specific start or end state (Figure 10.1). For example, most maps are held with North at the top, so when travel on a western road requires a northward turn, the map viewer must note that this will be a right turn, even though the map indicates "up."

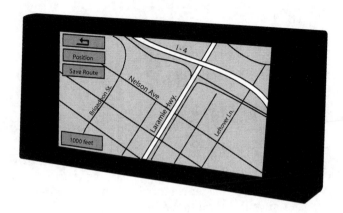

Figure 10.1 Example of an allocentric view.

These spatial demands of paper maps can overload older users, particularly in time-limited situations (such as driving in traffic in an unfamiliar area). Often navigation systems allow the same view, only with less of the map visible at a time due to the size of the screen. The benefit of allocentric systems is that most users are familiar with maps and can draw on previous knowledge. After all, increased knowledge is one of the benefits of aging. However, the spatial requirements of allocentric views can negate this benefit.

The second type of view, egocentric, can only be shown via a computerized navigation system. In an egocentric view, or "first person view," the user sees a similar view to the one out the windshield, composed of symbols. Streets and turns are still represented as lines, but the display has perspective (Figure 10.2).

The benefit of this view is that it reduces the need for spatial ability: a right turn will always be a right turn, no matter what direction the car faces. However, the view of the user is limited to their immediate vicinity and gives the navigation system control of the route. In an allocentric view, users may see a better, faster, or safer way to travel than the automated route chosen by the system. Users in an egocentric view will not have this option because they are limited to the area immediately around them.

The last view, the hover view, is also sometimes called the "third person" view. It is a mixture of allocentric and egocentric in that it gives an egocentric view, but it is as though the camera is above and behind the user. This allows a slightly allocentric view, such that more land and streets are viewable in the distance, but the egocentric benefit of reduced spatial processing requirements. Figure 10.3 provides an example of the hover view, with an accompanying diagram of the concept. The hover

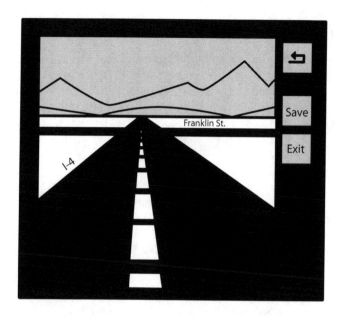

Figure 10.2 Egocentric view.

view still does not provide as much of an overview as the allocentric view, but allows more planning than the "just in time" egocentric view.

In summary, all three views have costs and benefits. The benefit of the allocentric view is similarity to the familiar paper map: the egocentric view requires no translation to understand what is approaching in the distance; and the hover view combines the two to provide more information, but does not look the same as the view out the windshield.

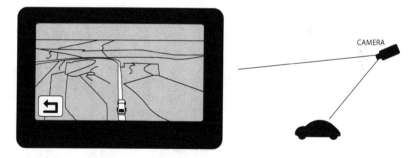

Figure 10.3 Hover view of upcoming turns and landscape.

10.2   Perceptual concerns

Navigation systems often have both visual and auditory displays. In some cases, this redundancy benefits the user, particularly when designers take special care to match the type of display to the task the user performs. For older users, there are particulars of these displays that need to be considered and tested. One issue is color: does the system show the route with a pleasing blue, perhaps a color that tested well with younger users? Blue might not differentiate the route enough from the background for an older user.

The resolution of small screens, matched with a high number of color indicators on a low-contrast background, can be a nightmare for older users. Indicate controls and symbols of high importance with high-contrast graphics. Make sure that important words are high contrast, relative to the background, and have a visual angle of at least 20 degrees when viewed from seat distance.

For auditory displays, follow the same rules outlined in Chapter 3 (Hearing) for phone menu systems. Use a naturally recorded voice rather than a word-by-word computer generated one. Because of the demands of the driving task, the sentence construction of audio displays for navigation systems is particularly crucial. The issues in auditory navigation design are summed up in these guidelines by Baldwin (2002), who specializes in auditory displays for older users.

In summary, there are a number of perceptual issues that accompany the smaller screens used for navigational displays in automobiles. These screens may be of differing sizes, and they are used in many environments, from the very bright to the very dark. We provide the following guidelines for improving perception of a display.

10.2.1   Guidelines

- Auditory information complements the visual and motor task of driving. When well designed, it can greatly improve older adult interaction with a system. It is especially useful for navigational assistance and alarms.
- A rule of thumb is to increase volume for navigational information to be 10 dB above levels used for younger drivers.
- Further, this level should be from +6 dB to +15 dB (signal-to-noise ratio, S/N) above background noise levels. These sound levels should be tested with users in multitask scenarios.
- Increase the context surrounding a verbal instruction. The order of presented information is more crucial to the older user than the

younger user. Provide information that requires reaction before additional context information.

- Although natural language in general is beneficial to older user comprehension, adding noninformative information harms comprehension. Adding context does not mean creating sentences. Often a list of choices or instructions is more understandable than sentence form.
- Keep the display consistent across tasks. For example, use the same display elements for all menus.
- Do not have more than three instructions in a message. An example of three content items in a message is: "Left on Main Street in one-quarter mile." Three items has been shown to be the maximum for younger users. We do not know for certain that older users need fewer content items, but it is probably a good idea to reduce them to one or two.
- Use recorded voices whenever possible, and avoid compression.
- Consider the potentially reduced reaction time of the driver and give navigational instructions with more time before action is necessary. Younger adults require about 300 meters advance notice for a turn. Older drivers need more advance notice, but be cautious of misleading them into turns before the actual desired turn.
- Keep contrast high for important display elements.
- Choose colors carefully, based on usability and aesthetics, rather than aesthetics alone.

10.3   *Movement control and input devices*

One of the main difficulties with car navigation systems is the proprietary input controls created by each manufacturer. Because they can be so different, from touch screens to side buttons to rocker knobs, use of each requires learning the input controls as well as the system concept. In the example for this chapter, the original interface was controlled by a knob that could be turned, pressed, and pushed in four directions. This type of control can be well matched to the design of the interface, provided it is arranged with buttons or other controls that follow the movement of the knob. Two issues to watch for that were present in the system redesigned here were mode errors and accidental activation. The multipurpose knob may perform many functions; however turning the knob operated a different portion of the screen than did "rocking" it left or right. Movement left or right may occur unintentionally while turning, putting the user into a different operational mode (mode error). The user may not notice this happened, leaving him or her frustrated at the undesired change in the display (accidental activation).

10.4   Usability assessment

An example of usability assessment for a car navigation system follows the general pattern of most usability testing with the addition of considering the particular needs and issues most likely to appear for older users. In the current example, we analyzed the most relevant issues for older adults, then developed a persona and user description to guide our thinking. Finally, we did a detailed task analysis of common functions of the navigation system and carefully analyzed each step in terms of the perceptual and cognitive demands on the user when used in an automotive environment.

Older users of navigation systems need to be tested in an environment similar to the one in which the system will be used. Age-related declines in perception and cognition are exacerbated by stress, multitasking, or other resource requirements. Thus, the results from interaction with a system in a quiet, single-task environment will not provide an accurate representation of the true usability of the system. This is true for the tones and auditory information (it should be tested in an environment with background noise), for the motor requirements (reach and pressure demands), and visual information (visual angle, glare, day and night lighting should all be considered). Also, usability testing should require the older user to perform tasks while some attentional resources are devoted to other in-car tasks.

10.4.1   User profile and persona

User profile information would be gathered via surveys and interviews to determine the common goals desired by older users, their frustrations with current systems, and demographics information about their backgrounds, physical abilities, and cognitive abilities. These data can be used to create a persona of a user to direct the thinking of designers and usability testers.

10.4.2   Personal description (Figure 10.4)

- 72 years old.
- Lives in Ann Arbor, Michigan.
- Children live in Michigan, New Mexico, and Wisconsin.
- David and his wife Pat (73), often travel together by car to visit children and now grandchildren.

10.4.3   Technological experience

- David drives a 4-year-old sedan with an in-dash navigation system. He has to have different U.S. areas loaded into the navigation system by the dealership when he travels more than 800 miles from home.

Figure 10.4 David Hong: retired high school social studies teacher, frequent traveler, grandfather.

- His main tasks with the navigation system include entering the addresses of his children across the United States and finding hotels along the way and near those addresses.
- His navigation system contains an autocomplete feature where the system removes letters that are no longer possible in the current entry. For example, if he types "balt," then letters such as "x" and "v" disappear because they cannot follow "t."
- He owns a cell phone that also has mapping capability but does not like the controls or the small screen. He uses the in-car system even though the cell phone would not require a trip to the dealer for map changes.

10.4.4 Tasks

David and his wife almost never use their navigation system when driving around Ann Arbor, Michigan (their hometown). However, they have come to depend on the navigation system for longer trips to see family. They most commonly enter a destination while they are in Ann Arbor and let the system guide them. These are multiday trips and often involve finding hotels along the way and entering a final hotel in the destination city. This means they rarely get to use any saved destinations and always need to use the text entry feature of the navigation system. David is goal-oriented when it comes to his navigation system; he would like to use it quickly with minimal letter entry even though the autocomplete feature never fails to surprise him and make him hesitate.

10.4.5   *Motivations, attitudes, and current problems*

Inputting the final address in his navigation system takes him a long time and is fraught with errors. Many of these errors come from moving the data entry knob in a direction when he only meant to turn it clockwise or counterclockwise. He gets very frustrated and at times does not use the system if his wife will be in the car with a map. The fact that he is usually comfortable with technology that has a keyboard makes him more frustrated with this system because it slows him down. Although there are some voice inputs, the system often does not correctly understand what he said, and he does not know how to correct errors made by the system. Last, David has successfully navigated all of his life and suspects that the system makes errors in efficiency; he believes he could have done better. David's desire for speed in his text entry is a major source of his frustration with the system.

10.4.6   *Scenario and task analysis*

David and his wife are planning a long road trip to visit for a week with their daughter and new grandson. Because their daughter recently moved, they need to input a new city, state, and address for her. This involves using the text entry portion of the display.

Table 10.1 is an example task analysis and proposed redesign of a sample car navigation system, specifically the task of entering the user's destination via text. Though the task analysis revealed many issues applicable to all users, we were careful to consistently consider the needs of older users in the recommendations section for each screen. Each step in the task flow for the analysis is named and contains an illustration of the navigation system. Our observations regarding that step in the task flow and display screen follow, pointing out potential usability issues. We follow those observations with a redesigned display screen and a list of recommendations that were taken into account when redesigning the display. As a disclaimer, the redesigns we provide are meant only to illustrate one possible solution.

10.5   *Specific design changes/recommendations*

Perceptual
- Increase perceptibility.
  - Through visual enhancement
  - Through auditory enhancement

Cognitive
- The allocentric view is good for planning and giving an overall concept of travel to a destination. It should be used initially to allow users to understand the route they have chosen and make changes if necessary (not depend on the automation of the system).

Table 10.1 Task Flow: Enter a Destination into an Automobile Navigation System via Text

Step 1	*Recommendations*

Access the Menu. Press the menu button.

Figure 10.5a Original screen.

Observations

- Use of gray decreases the contrast on the screen.
- Icons are difficult to link to functions. For example, the intersection looks like a button, and Previous Destinations suggests it will return the user to where they came from.
- Choice indication is minimal: a thin colored line around current choice and a colored icon. Current color choice is blue, which is harder to detect by older adults.
- Control knob is a well-matched device to this interface, but does not need representation on screen (the circle in the center) unless it helps to indicate the current choice.
- Unavailable choices are grayed out, a good way to indicate unavailability, but the reduced contrast may make them unreadable.

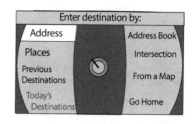

Figure 10.5b Redesigned screen.

- "Brushed Aluminum" look has been retained, but darkened for background.
- Icons are removed.
- Knob has additional orientation information added to augment current choice.
- Current Choice indication is augmented with long wave color (Orange) and increased area.
- Unavailable choices are grayed out, and background of button also serves as an indicator.

Step 2

Select "Address." The display changes to:

Figure 10.6a Original screen.

Observations

- Communication does not follow an expected order. "Find address by: in GEORGIA; city or street; press down to change state from Georgia."

(continued)

Table 10.1 Task Flow: Enter a Destination into an Automobile Navigation System via Text (Continued)

- Toggle button may create accidental activation for users with even slight movement control issues.
- Current State ("Georgia") is indicated in textbox. To change destination state, press dial down.
- Manual recommends entering street before City when entering text.
- "CHANGE STATE" control violates the proximity compatibility principle: controls should be located near their representation on a display.

Step 3

Selecting the City. Current city is displayed in center. To change destination city, press dial down

Figure 10.7a Original screen.

Recommendations

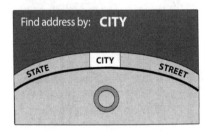

Figure 10.6b Redesigned screen.

- Screen has been rearranged to place control notation near the display item control. The order of steps is preserved from the original design; first choose state, then city or street.
- CITY has been made to look like a choose-able button.
- In essence, there are now three buttons on this screen: State, City, and Street.

Observations

- On previous screens in the original design, current location was indicated via a textbox. On this screen it is indicated via a low-contrast text-on-background, showing inconsistency between screens.
- What does current city indicate? The city one is currently in, or the city displayed on the background above?
- Previous screens changed the location by pressing down on the dial. Now, on a conceptually similar screen, the same action produces a different result: "Change to spell mode."
- All-capital letters are difficult to read.

Table 10.1 Task Flow: Enter a Destination into an Automobile Navigation System via Text (Continued)

Recommendations

Figure 10.7b Redesigned screen.

- Button now says what the exact action of the button will do: "Select Mobile, AL."
- "Change to spell mode" is now a button and states what action will occur: "Spell city name." Also, it has been moved closer to the text field it controls and is in a larger, higher-contrast font.
- All-capital font is changed to sentence case.

Step 4

Text entry screen for city name.

Figure 10.8a Original screen.

Observations

- It is not clear how to access numbers and symbols at bottom.
- This screen demands a difference between turning the toggle knob and rocking it left or right. Rocking it will result in deletion of a letter or a screen change to "List" format.

- Turning the knob will choose letters. The turn maps well to the choice of letters along a curved display. However, it does require control not to toggle a direction while turning.
- Good indicator for currently chosen letter.
- "Enter" symbol may not be known by older users. Is this enter different than the other screens that use a toggle button push?

Recommendations

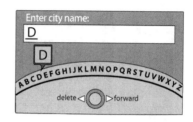

Figure 10.8b Redesigned screen.

- Numbers at bottom are removed.
- "Hits" are removed.
- List of cities is a choice not a direction.
- Enter symbol is removed from alphabet.
- Contrast increased.

(continued)

Table 10.1 Task Flow: Enter a Destination into an Automobile Navigation System via Text (Continued)

Step 5	Step 6
ERROR STATE: System does not recognize audio. Presents choices for oral confirmation.	By Toggle Button: Select City, and the display then changes to the Enter city name screen. The system will display a list of city names, with the closest match to the name you entered at the top of the list. Select the number (1–6) of the desired city from the list. Asterisk represents a poorly mapped city.

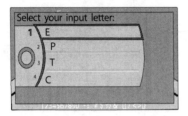

Figure 10.9a Original screen.

Observations

- Letters are mapped to numbers. It is unclear if they are ordered by likelihood or how many choices would be available in an extreme case on this screen.

Recommendations

Figure 10.9b Redesigned screen.

- Letters are no longer mapped to numbers. They are arranged in alphabetical order.

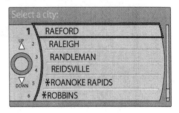

Figure 10.10a Original screen.

Observations

- Again, the system creates a numerical mapping for city names.
- "Up" and "Down" are written on the toggle button for the first time. Is this because the typical way of choosing from a list (turning) is disabled?
- Scroll bar as an indicator is small and low contrast.

Recommendations

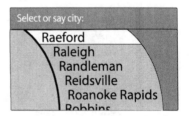

Figure 10.10b Redesigned screen.

- Available cities are arranged so it is clear a turn of the knob can select them.

Table 10.1 Task Flow: Enter a Destination into an Automobile Navigation System via Text (Continued)

- Scroll bar has been removed, and instead the bottom city name is "partially below the fold" to indicate more cities present below.

Step 7

Use the toggle button to enter the name of the street. Press Enter.

Figure 10.11a Original screen.

Observations

- Numbers are accessible, though instructions are to enter the Street Name.

Recommendations

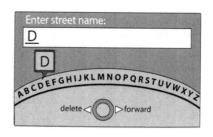

Figure 10.11b Redesigned screen.

- Numbers at bottom are removed.
- "Hits" are removed.
- List of cities is a choice not a direction.
- Enter symbol is removed from alphabet.
- Contrast increased.
- All auditory responses increased by 16 dB, as recommended by Coren (1994).

- The egocentric view is most clearly mapped to the view of the user; however, it is limited in scope and can be a crutch the user cannot function without.
- The hover view is a good compromise and should be used during navigation to reduce the need for fluid abilities when they are already being taxed by driving.
- Simplify the feature set, or hide advanced features.
- Decrease complexity.
  - Through shallow menus
  - By displaying choices
  - With environmental support

Movement

- Match the choice of input to the design of the screen and on-screen controls, such as buttons.
- Specifically test older users under multitask and movement scenarios to check for mode errors and accidental activation.

10.6   Summary

A car navigation system is a great example of a complex display made more complex by the environment in which it is used. Designing such a display should focus not only on the perceptual attributes of the display, but especially on the cognitive effort required to understand and utilize the display. A large font may be the first step in readability, but understanding depends on interface cues such as flow through a dialog and the interaction of the input device and on-screen controls. The best designs begin with keeping in mind the perceptual and cognitive capabilities and limitations of the user population and end with thorough user testing with members of that population.

10.7   Suggested readings

Baldwin, C. L. (2002). Designing in-vehicle technologies for older drivers: Application of sensory-cognitive interaction theory. *Theoretical Issues in Ergonomics Science,* 3(4), 307–329.

Coren, S. (1994). Most comfortable listening level as a function of age. *Ergonomics,* 37, 1279–1274.

Kern, D., Schmidt, A., Arnsmann, J., Appelmann, T., Pararasasegaran, N., and Piepiera, B. (2009). Writing to your car: Handwritten text input while driving. *CHI 2009,* April, 2009, Boston, MA.

Llaneras, R. E. and Singer, J. P. (2002). *In-Vehicle Navigation Systems: Interface Characteristics and Industry Trends.* Driving Assessment 2003: 2nd International Driving Symposium on Human Factors in Driver Assessment, Training and Vehicle Design.

Tsimhoni, O., Smith, D., and Green, P. (2004). Address entry while driving: Speech recognition versus a touch-screen keyboard. *Human Factors, 46*(4), 600–610.

chapter eleven

Conclusion

Technology is a well-accepted part of daily life. Fortunately or unfortunately, it is also well accepted that there is a constant torrent of new and changing technology. Just in the past decade we have witnessed the rapid proliferation of mobile technology, consumer electronics, and the internet and the web into almost every aspect of our lives (recreation and entertainment, learning, healthcare, work). Even within each category of technology there is explosive growth and change that may affect users' expectations and behavior. Take, for example, the shift from desktop applications (e.g., installed e-mail clients, word processors) to web-based applications (Chapter 4). These new web-based applications can appear identical to desktop applications but behave in unexpected ways to the user. While technology trends and interface paradigms will change, fundamental patterns of age-related change in perception, cognition, and motor control change slowly if at all. It is for this reason that we focused this book on explicating the precise ways in which older users vary along these dimensions.

One counterpoint to understanding why older adults might have difficulty with specific technology is that the "new" version of the system or a competing brand has removed the issue—problem solved! This misses the point of research on basic perception and cognition—just because a single technology has been improved does not mean the mistakes from the first design will not occur in future technologies. Knowledge of these patterns of change will be applicable even as the specific examples of technology (and displays/user interfaces) change.

We began our introduction with a general overview of usability for aging and evidence against some common stereotypes of aging. Older users want to use technology, especially when they understand its benefit. If they are thwarted in technology use, it is often due to poor interfaces and lack of perceived benefit rather than inability to learn or an unwillingness to adopt new technology. The next few chapters were what we termed the "fundamentals" chapters, each providing an overview of the research literature on cognitive/perceptual change. In these chapters we sought to inform designers and usability experience professionals about some of the lesser-known changes that tend to come with age, such as contrast sensitivity (Vision), need for prosody (Hearing), fluid versus crystallized abilities (Cognition), and the Hick-Hyman law (Movement).

Along with descriptions of the more well-known changes, these chapters provide a solid base from which to consider new designs. In conclusion, it is not difficult to develop a basic understanding of age-related change, but it requires continuous attention to the possible effects of these changes as well as their interactions when a display requires more than a single cognitive or perceptual ability.

Even designers, researchers, and usability practitioners who have a good understanding of the fundamentals of age-related change need to test their designs. In Chapter 6 (Older Adults in the User-Centered Design Process), we outlined how to include aging considerations in the basic usability process. Though most techniques developed with younger users transfer well to testing older users, a few additional considerations for aging participants can make usability testing more valuable. The strongest message to come from this chapter is to include older users in design and testing.

Finally, we devoted the third portion of the book to chapters on redesigning specific systems using knowledge from the fundamentals chapters as well as the chapter on testing older users. In these chapters we took generic examples from commercial displays and walked through the steps of a usability analysis. Also included in these chapters was a general discussion of the challenges each type of system pose for older users of those displays. Although we did not perform usability tests with actual users, this would be an important next step for any practitioner following our recommendations and is built in to the user-centered design process (iteration of design and evaluation).

11.1   Themes

Looking back at the chapters, there are several recurring themes that ran across chapters and were not tied to any particular ability, change due to aging, or display type. These themes were our take-home messages in this book.

11.1.1   Aging is associated losses, gains, and stability

One key idea we hope the reader takes away from this book is the idea that aging is not indicative of *general decline* in perceptual, mental, and physical abilities. Popular stereotypes of aging seem to focus on the declines, but what is less well known is that these declines are somewhat balanced by gains in other faculties or even what does not change with age. As discussed in Chapter 4, fluid abilities consistently show age-related declines, but those losses are offset by gains in crystallized intelligence. The preceding chapters discussed the nature of the gains and losses because, conceptually, interfaces should be designed to take advantage of the "gains" while minimizing demands on the "losses." More concretely, we showed only one possible way to do this (using alternate forms of information presentation, such as the tag-based interfaces in Chapter 4) but there may be other creative ways to accomplish this goal. Some abilities are more stable than one might expect, such as continuing expert motor skills or accuracy of touching a target with the hand.

11.1.2   Age-sensitive design requires knowing the user

There is immensely wide variability in the abilities and limitations of older users, even more so than for younger users. There is no single "older user"; the population of older users of technology may be more heterogeneous in abilities, knowledge, motivations, and lifestyle than users of other age groups. As repeated in previous chapters, older users vary tremendously in their perceptual, cognitive, and motor capabilities in addition to the level of knowledge and skills. These users differ not only from each other but also differ from younger cohorts. We hope that the previous chapters have shown some of the critical and, in some cases, subtle ways that older users vary from users of other age groups. For example, preattentive visual search (noticing conspicuous things in the interface; Chapter 2) shows little to no age-related differences, and Chapter 5 illustrated that there are little to no age-related differences in simple reaction time (when users must simply press a button), but sizeable age-related differences in choice reaction time (when users must decide which button to push).

In addition, documenting the different motivations for using and attitudes about technology can inform the design of technology, because usage patterns are likely to be different. For example, there is currently increased interest in personal location tracking and sharing such that a user can broadcast their current location to their friends. Older adults may be interested in this technology for different reasons: to track friends or spouses who tend to wander due to Alzheimer's or to serve as a safety net for themselves in their travels. And, of course, older users may also wish to broadcast their locations to friends to engage socially! The underlying technology may be identical (GPS, mobile phone technology, web-based maps), but the display will likely be drastically different based on motivations and usage patterns.

11.1.3   The usability barrier of technology adoption

It is a stereotype that older adults do not want to use new technology. Instead, they (like people of other age groups) may be more judicious and weigh the costs of adopting new technology (in time and costs) against the benefits. One major category of costs is usability. Perceived usability has been shown to be a major determinant of whether older adults choose to adopt new technology.

11.1.4   Usability affects older adults' adoption and usage of technology

Whether someone chooses to adopt new technology can be complex. Factors such as cost, usability, functionality, and needs dictate whether some technologies are adopted. When technologies are perceived to be usable and better than existing technologies, they will adopt. An anecdotal illustration of this is the incredibly wide adoption of the Wii video game system by older users. Prior to the Wii, the older adult segment was not known to be a rapid adopter of video games, but the Wii is becoming a fixture in senior centers and independent living communities. The popularity of the Wii among older adults can be partially attributed to the nearly effortless interface design (primary controls are hand and arm motions, not buttons) and the benefits of use (exercise, social interaction, and fun).

11.2   Important future goals

As of 2007, about 12% of the population was over age 65 (U.S. Census Bureau). Older adults are expected to comprise over 19% of the population by 2030 (U.S. Census Bureau). Technology and the displays required for this technology are already heavily integrated in daily life, from cell phones to the internet, and this integration increases each year. For example, many commercial websites no longer allow phone contact or provide phone numbers; communication must take place through an electronic medium. Although there may be cohort effects, where the older users of tomorrow are the technologically savvy users of today, no one is immune to age-related change. Taking into account these potential changes is a worthwhile way to make displays usable and desirable to an older population.

11.3   Concluding remarks

The user experience analyst or usability engineer is often tasked with "making things usable" in short time periods with few resources. We intended this book to be a convenient source of information relevant to the design of displays. The goals of this book were to provide a primer of age-related changes in older adults' capabilities and limitations that are relevant to the use of displays so that designers would have a thorough and practical reference. We also sought to provide conceptual application—a translation—of these basic facts in the design of displays. It is our belief that seeing this translation in action is one of the best ways to learn what to look for in displays to make them more aging-friendly. We hope the reader has come away with new information about aging and the process of designing displays for older users.

Index